OSAU-3 Presents

What, When, Where, Why, How

and

WHO IS US?

an

AWTbook™

by

H. FRANK GAERTNER

WORKBOOK PRESS LLC
187 E Warm Springs Rd
Suite B285 Las Vegas NV 89119 USA

Website: https://workbookpress.com/
Hotline: 1-888-818-4856
Email: admin@workbookpress.com

Ordering Information:
Quantity sales. Special discounts are available on quantity purchases by corporations, associations, and others. For details, contact the publisher at the address above.

ISBN-13:   978-1-961845-01-5   Paperback Version
                  978-1-961845-11-4   Digital Version

PUB. DATE: 01/17/2024

OSAU-3 Presents
What, When, Where, Why, How
and

Who is Us?

an

AWTbook™
by
H. Frank Gaertner

Table of Contents

Pages

Dedication

OSAU-3 would not have been written without the inspiration and guidance I received many years ago from Dr's William O'Neil (Ventura-College, Bacteriology & Botany-1959), Adelaide Evenson (University-of-Arizona, Mycology-1961) and Henry Koffler (Purdue University, Molecular Genetics, 1964). Hence, it's a great privilege for me to be able to acknowledge their life-long dedication to the physical and natural sciences. The following referenced-video summarizes that dedication with a message I know each would want me to share. Please check it out *after* you read OSAU-3.

YouTube *"A Life on Our Planet"*, David Attenborough, Wild Life Focus (6:20:59).

I think Attenborough's countdown and description of the impact we humans have unleashed on the *essential* biodiversity of our planet is without doubt *the most important account of its kind.* The audible version is narrated by a great story teller, the author himself. The book's visual content is also beautifully presented on *YouTube.*

My literary-foil and I are most grateful for Quora's presenters, especially Viktor Toth, and for the YouTube-presenters, especially Lex Fridman, Jim Al-Khalili, Arvin Ash, Derek Muller, and Kathy Joseph. But my foil and I specifically dedicate *"Who is Us?"* to all those who can use our book's authorship method. Why so dedicate? Our own beta-test shows just how well the method works. We found it easy to write this complex, multifaceted book that finds us *Us's* dangerously confined to a spinning, tiny, blue dot endlessly entertain*ed* by a vast Self-Assembling Universe.

Foreword: How to Use an AWTbook™

The book you are about to read is different. It's a *"voice-activated, smartphone-in-hand", AWTbook™,* a new form of literature that's easily recognized by its exclusive use of interactive Audio, AI, Website and YouTube references. Such audio/AI/visual references can bring instant comprehension to difficult and otherwise confusing subject-matter. For a quick example of an *AWTbook™* in action, grab your smartphone right now to watch the following six-minute video. You can do that just by "waking up" your phone with a *"Hey Siri, Google, Alexa or Bixby"* followed immediately by a second, voice-activating command such as, *"YouTube! Your Body's Molecular Machines".* Now, if you are not hugely *impressed-intrigued-enlightened,* having seen what I'm talking about, you can simply move on to seek a more standard form of literature. On the other hand, if you are amazed by how such a new-age, smartphone addition to a standard book can light-up its content, I think I can safely say that you have become a fan. In addition, depending on your level of enthrallment, you might want to introduce yourself to my new, career-inspiring literary-art-form by perusing the first *AWTbook™, OSAU-2.*

Goodreads: Book Reviews, *"Our Self-Assembling Universe, OSAU-2 an AWTbook™",* H. Frank Gaertner (81 pages). *FYI: OSAU-1 has a bluish, red-lettered cover. OSAU-2 looks similar but has a yellowish, sparkly gold-lettered cover. Both have extensive 5-star reviews but OSAU-2 is a much better value since it includes all of OSAU-1, corrected to eliminate a few important errors, has a wonderfully mind-boggling quantum-preface, and stands as the first publication of an AWTbook™.*

To further entice your reading of the current book I've included below three Arvin Ash, YouTube references that one can voice-activate to show just how an *AWTbook™* enhances text with state-of-the-art, audio/visual presentation. Also these three do a good job of illustrating the type of content one can expect as they read this book. In particular, be sure to watch the third video that highlights the magical carbon atom *and its role in the origin and assembly of all things living, including the Us of Who is Us.*

Notice: Please see the running-time designations at the end of each video-reference. Also take heed of the JL, LW, B designations that I've added at the end of each reference.

I've added these to let you know my preferred method for efficient use of each video's information. For example, some of the video-references do not need to be watched. To these references I've given the JL designation for "JustListen". Other references should be "ListenWatch". These get the LW designation. Those that I think might need a JL as well as an LW I've assigned the letter B for Both. The B's are the most important videos and therefore can use special attention. I find walkjogrunning-JustListens followed by sitdown-ListenWatchings are great ways to get important concepts across and mentally fixed. Why add these designations? I don't like to waste time and I'm guessing you don't either. I like to walkjogrun as I justlisten (JL). Justlistening as I walkjogrun has kept me active and in good shape for years, and still does so at my current age of 85. I take off most mornings at about the same time for a one to two hour walkjogrun. It's so motivating to know that I've got something cool to listen to as I exercise. Therefore you, too, might want to JL as you walkjogrun with all of my video references, and then do an LW on the ones that you've especially enjoyed while doing a JL. Having done so, one knows for sure that you have a particular desire to do an LW. Some of the videos are simply lectures that have no special visual aids. Obviously, these only require a JL. Others may need a B or a special LW as you read this book just as I've done with the ones I've so designated with B's and an LW.

YouTube *"Unbelievable: Learn Nuclear Physics in under 12 Minutes"*, Arvin Ash (12:28-B). YouTube *"Why does Light Exist? What is its Purpose?"*, Arvin Ash (15:0-LW).

YouTube *"Why is all Life Carbon Based, Not Silicon? Three Startling Reasons!"*, Arvin Ash (14:04-B).

FYI: To collect the videos for JL's as I *walkjogrun*, I simply add the reference links to my phone by activating the videos with the type of commands that I've described above. I then use the *shared-option* that always shows below each YouTube video to share the video's hyperlink to my own email. Then it is easy during my *walkjogrun* to call up the videos I want to JL by clicking on the link that's been titled and saved in my email.

Also, it just occurred to me that it might be possible to do an LW as one *walkjogruns* using a VR headset. Maybe so, but it might be dangerous depending on where one is going. Tech is zooming so fast, I have to ask, have I already been antiquated?

Preface

This third volume of *Our Self-Assembling Universe*, *OSAU-3* for short, continues where *OSAU-2* left off. In *OSAU-2*, the first *AWTbook™*, I included the entirety of *OSAU-1* primarily to emphasize an important point that I had raised which, apparently, nobody else had previously noted. In that first 2015 edition I stated that apparently- overlooked point as a question. *"All material things, including all life-forms, are not just made of atoms, are they?"* Naively, I thought everyone would suddenly realize that they have been misdirected, much as a magician does when they misdirect their audience. But, if the magician or another person should happen to explain the trick, suddenly everybody understands the truth and says, "Clever, but how could I be so stupid!" Thus, I thought everyone would similarly get it when I explained what I thought to be a misdirection that was staring us in the face. I explained, *"Of course all material things are not just made of atoms, are they? All such things are made by them."* Now you get it, if you haven't before. You, too, see the trick and it's obvious, right? Nevertheless, my little replacement of the word *of* with *by* must be hard to swallow because, aside from my own use of the word *by* in the atomic self-assembling context, we continue to be misled by the same solidly-ingrained misdirecting phrase that misled me. As of this writing, I'm pretty sure that all you will ever hear, except from me and possibly a few others, is "everything is made *of* atoms", and that misleading word *of* is used even if the statement is coming from the particle physicists that I think should know better. Moreover, until the phrase *made of atoms* is replaced with *made by atoms,* it's likely one will only hear in *this context the word "of" instead* of *"by".* And this is true even though it has now been well documented that *atoms on their own* assemble every massive thing in Our Mathematical Universe! But, but, but! What about God? Okay, okay, already. You're right. A Supreme Simulator probably is doing it with help from algorithms that determine the rules of thermodynamics, electrodynamics, random motion etc.

Maybe I have missed something but I believe my suggested change in that prepositional phrase is so important and correct I'm willing to confront any particle physicist who thinks otherwise. I believe this is an oversight and until it is challenged, the importance of using *by* in this context will continue to be universally ignored. Why does it matter? Why am I so adamant about this change in wording? Because the use of *"of"* rather than *"by"* totally denies the active, fantastic, magical and yet very real accomplishments of the approximate 7,000,000,000,000,000,000,000,000,000-member atomic-workforce that re-assembles much of our fragile human frames into existence each and every day! Moreover, that inanimate atomic-workforce doesn't just reassemble our frames. As one reads this book, your atoms are at work. Among many other marvels, you will discover how our atoms create armies to defend us, *cellular* energy generating batteries to charge us, movers to move us, menders to mend us, membrane ion-channels to jolt us and, above all, the cells called neurons that bring us into the baffling world of electrically-coded conscious existence! This all may sound like I am trying to be anthropomorphic. I am not. We are multicellular creatures made *BY* atoms! That's simply a fact.

Oh yes, I think you can now understand why I am so perturbed and why I am so determined to be undaunted in the face of the science community's silence. Therefore, just to be totally clear about it, we are without doubt made *of* atoms but, also without ANY doubt, *we are made by the very same atoms of which we are made!* I rest my case with the following two video references. The first one is once again from Arvin Ash, one of my most favorite YouTube-presenters on the subject of particle-physics. Here, Ash beautifully summarizes and supports all that I've said up to this point about atoms. However, please note, not once does Ash use the active preposition, by, when he talks about all of the unbelievable things that atoms do. The second video-reference links to a fantastic, science-documentary presenter. He also does a great job of supporting my contentious contention. But, once again, notice the title of his documentary!

YouTube *"Why do Atoms form Molecules? The quantum physics of chemical bonds explained"*, Arvin Ash (13:24-B).

YouTube *"Why is Everything Made of Atoms?"*, History of the Universe (45:41-B).

I had hoped to bring more attention to my *of-to-by replacement-contention* when I published *OSAU-2* in 2022, wherein, one is introduced to the effects of quantum mechanics and how this disruptive, relatively new science totally supports the proper use of *"by"* when one speaks of atoms. And, more importantly, how that little change in the preposition *of to by totally re-imagines* all of the remarkable things atoms do as they Self-Assemble *Us* and the rest of Our Universe. In addition, my *OSAU-2* expose'

WHO IS US? An AWTbook™

took *OSAU-1's unappreciated, completely ignored assertion* to an entirely new level. Therein, I included Paul Dirac's 1928 equation that took his already *"hard to swallow"* ideas to a realm that most thought to be *"impossible-to-believe"*. But here's the thing, it turns out that some of the Dirac equation's most outrageous, *impossible-to-believe* predictions have already made their way into modern medicine! For example, have we all not heard of *PET-scans* and *antiparticles?* If not, know this. The P in PET-scan stands for *positron*. Positrons are positively-charged, *antiparticle electrons*! It's the recording of the flashes of light that occur when positrons and electrons meet, *only to annihilate each other,* that make the remarkable PET-scan images possible. When Dirac published his equation in his one-upping of Einstein, the equation looked beautiful but there was a problem. The equation predicted the *existence of antiparticles!* And at the time Dirac published his equation, most thought him to be crazy. Antiparticles weren't thought to be possible. They didn't exist! But, obviously, they do. Interesting, isn't it?

YouTube "What is PET/CT and how does it work?", Siemens Healthineers (3:53-LW).

Nevertheless, one might still wonder about Dirac's equation and shake their head with a "No way. I still can't believe it!" How so you might ask? Why would anyone have a problem believing this, now that we know for sure that antiparticles exist? Why? *Because* Dirac and his equation also have this to say, and I paraphrase; *Everything has been and continues to be self-assembled by an atomic world that manifests itself from empty space. Seething seas (fields) of virtual particles and antiparticles materialize matter by flashing in and out of existence from apparent nothingness!* Or, as I like to rephrase my own paraphrase, *vast co-ops of ingenious, self-assembling, atomic-class universe-builders are summoned from empty space to perform an unbelievable task, the invention and assemblage of everything we hold true and dear.* That might be a little hard to take but just try grappling with this thought for a minute. It means that you, me and the astonishing surface of the planet on which we live are not just made *"by"* atoms. *Everything* is made *"by"* atoms that manifest themselves from *electrons* and from things particle-physicists call *quarks "materializing" from electromagnetic fields of excitation that occur in empty space*! However, if you happen to be one who has no trouble imagining vectored fields of annihilating virtual particles and antiparticles that flash in and out of existence in empty space to make themselves into something, you have much better imagination than I do. Maybe the following video- reference-link will help those of us with weak imaginations. It certainly helped me.

YouTube *"The Crazy Mass-Giving Mechanism of the Higgs Field Simplified"*, Arvin Ash (13:03-LW).

At the conclusion of this book I'll have more to say about how this mind-altering concept answers a lot about *Who is Us* and the arguments that some of us *Us's* have over Design, Intelligent or Otherwise. Also at the conclusion of this book I am adding a recent, detailed, YouTube video reference that summarizes in clear audio/visual presentation most everything you may have missed or want to review concerning the complex details of a Universe Assembling itself. Even though the narrator sounds a bit constipated, I like this video a lot. It's an accurate summary of OSAU-1. To see this summary check out one of the final references in my reference list: "Beyond the Atom: Incredible Plunge into the Heart of Matter".

However, before I go on and before I get too carried away getting us hackled up over all of our individual strange thoughts and rantings, I think it might be a good idea to reveal something that I know to be true about each of us and, specifically, let you all know a little about my unique self. Why on earth would I want to do that? I think it will help me explain why I think it's possible for us Us's to have a meeting of the minds over some very controversial subjects, one of which is the existence or not of Intelligent Design. I also think our coming together just might be the most critical step on our way to gaining a better understanding of Who Us Us's are. The following statements may at first sound shocking but you can calm down because I think you will eventually agree that they are perfectly innocent, very understandable statements. But I'll grant you, it might take a minute.

Here goes. This is undeniably true. We are all supernatural. And we are all supreme beings. How can I say that? First, there is nothing natural about our existence. Every word that follows in this book proves that point. Second, we are all supreme beings of sorts. To wit, any dogs, if they happen to be deep thinkers, know we are supreme beings. And, along with dogs etc., many, including me, think all of us multicellular creatures are conscious at some level and at some level, all of us are intelligent designers. But if you have trouble accepting how that statement could possibly be true for all creatures great and small, check out the amazing talents of the puffer fish and the outrageous behavior of slugs, which are featured below in the chapter called "All That Thrashing About".

Lastly, with regard to our getting together on the controversial Intelligent Designer subject, an obvious necessary step will be shrinking the gap between the Us's who have spiritual beliefs, the Us's who have simulation beliefs and the Us's in the world of ordinary folks and particle physicists who believe Our Universe can't be simulated because there ain't no Simulator. Coming together on such a controversial subject may sound impossible, but maybe the gap can be shrunk. However, for me to be of any help in that regard it occurred to me that I, Frank, may need to be totally Frank, especially, before I present the coming incontrovertible evidence and supportive experimental proofs for a Simulation and a Simulator. Oh yes. There's no doubt those experiments and the evidence that they will provide are coming. So one can get all heated up about it but I think everybody reading this book will know where I'm coming from, why I think simulation of matter is more than likely, and understand why I think strong evidence for the Simulation may already exist from particle physicists, the JWST and the growing number of scientists involved in immortality research. Stay tuned.

Having said all that, I will begin my own story and, just so you know, for me it's a very deep and personal love story. I'm pretty sure if you have yet to feel it, after having read this book you will gain evidence for and maybe even feel, if you don't already, the love-connection to which I refer. So, to be clear about it. There is no doubt about it, I love my life, my wife, my x-wife, my children, my grandchildren and my friends but even though I may be forced to love all of us Us's, I absolutely do not love what some of us Us's do. However, that love, or lack of it, is not exactly what I'm talking about. As this book will show, all creatures, the good, the bad, the ugly and the lovely have their part to play in this "boat" in which we find ourselves. And, because we are all in the same boat, somehow I think we'd best make that special, all inclusive "love" connection because we are all in this thing for life. I always shudder a little at this thought for, I submit, this life we are in might be "one without end". Why would I even think that? There is no denying it, there is only one reason I am not you, a dog, a plant, a mosquito, a microbe, or even some kind of space-alien located somewhere else in the vastness of Our Universe. That single reason, which is so certain, is just this. "I" simply can't be any other "you" because I am a "me"! But hold on. Maybe that's wrong. Maybe it will be possible for me to be a you at sometime in the not too far-distant future or maybe, even, join with another entity someplace else in this Universe? And if that's true, does that not mean somebody who is alive today might actually live long enough by "mind melding" with more than one "you" as a way to "live forever" inorder to be able to travel to the stars? That's surely off the charts but read on, for that's one of those future events that I think might be coming our way "relatively soon". Our world is changing rapidly. It looks like there might be a way for me to superimpose my brain with yours. Therefore, might it not be worth thinking about one's actions in this boat we're in while we are at it? Just saying. But enough of the futuristic conjecture. Here's my disclosure.

Disclosure

I am a retired scientist with a background in music, chemistry and molecular genetics. I am also an alcoholic who had his last drop the day after the total eclipse of the sun in Mazatlan, July 11, 1991. During that year I had three, very profound, life-changing experiences. The first occurred at three in the morning when I awoke from a fitful sleep to find myself audibly whispering in the dark, "Who are you?" With no hesitation I got an abrupt, distinctly audible, commanding response. "I'm love". Now I was really awake and more than just a little interested. No longer whispering, I responded. "How real are you anyway?" The no-nonsense answer shot back, "Just as real as you want to make me!" Was I hallucinating? Probably, but nothing even suggesting voices in my head has happened before or since. Initially I was jolted but the shock was instantly replaced by a peaceful state that soon left me in a blissful sleep. I awoke that morning feeling as though I'd had a very important other-worldly visit.

It's now thirty-two years later, the feeling has not left and I can say the "visit" definitely contributed to what happened next. Just a few weeks later my wife and I had a serious argument that sent us to counseling. Our astute and very effective counselor instantly and specifically knew exactly who we were, and therefore had no problem knowing who I was when she said one day, when she thought I was ready to hear it, "You know, Frank, I think you might be an alcoholic". Since everything she had said up to that point rang true, I didn't resist when she invited me to go with her to her own AA meeting. Fortunately, I was persuaded at the very first meeting to fill the groups vacant position of secretary-treasurer.

As it turned out, the position left me obliged to return every week where I was encouraged to honestly share just how badly I needed a drink. No question about it. By my own admission, I knew I was a genuine alcoholic. Sadly, even though I knew the addiction was killing me and knew that I really needed to quit, I couldn't. Try as I might AA hadn't worked. After months of abstinence I gave up. I couldn't stand it. I was in serious pain. I couldn't sleep and I couldn't stop obsessing. My predominate thought was, "I have to have a drink, I have to have a drink, etc., etc., etc."

So, after another obsessive, miserable, pain-filled night of no sleep, I gave up and headed for my liquor-cabinet that was still fully stocked with booze. I can still hear myself saying "You idiot. You must have a death wish!" One might think that thought would have stopped me but it didn't. I reached for one of my two bottles of rum, but Just as I had my hand on the bottle, I remembered what I'd promised to do at the AA meeting. If such a moment were to arise, I was supposed to stop myself, get on my knees and start praying for help. Aarrgghh! What nonsense! As if praying could help. Past events and my then scientific views of reality had convinced me that praying was ridiculous. However, I was desperate. Even though I knew it was total BS, I had promised and because I am the scientist that I am, I had to try the experiment, stupid or not. So I took my hand off the bottle and made my way to an empty, dark room. I knew I was in for it. I was supposed to get on my knees but there was nothing I was going to be able to do to force my stubborn self to do something so idiotic.

Pathetic, no? There I was, standing in the dark for the longest of time saying to myself over and over, "This is ridiculous!" But my stubborn scientific experimental cerebral hemisphere was even more stubborn than the dominant half of my brain and that kept me stuck there until I completed the damn experiment. In the end I finally forced myself to my knees even as the stupidity of the experiment worsened. How could it get worse? OMG, now I had to say, "God, please relieve me of this obsession!" Thank God for the suborn scientific half of my brain. I had to complete the experiment. Still, I remained frozen in place. I not only had to say that phrase, I had to say it out-loud while I was on my knees. I was thinking the words but I had been told I couldn't just think them. $%&^%!! At last after another mind-muttering, cursing, long time, I said, "Gah - - !!!" But that's all the further I got! I didn't even get the entire word God out of my mouth! At that instant it happened! And it happened just as I had heard it had for a few other alcoholics, including one of AA's two founders. I had been zapped by the alcoholic's miraculous bolt of spiritual lightening! I was immediately overwhelmed with such huge feelings of remorse, gratitude, joy and relief that my eyes filled with tears and my nose with buckets of snot. A severe pain in my head and throat, which I had been experiencing for weeks left, and I mean instantly! My desperate, obsessive need also instantly vanished and 32 years later it has never returned!

To say that again and make this long story much shorter, I now have 32, wonderful obsession-free years of total sobriety. Therefore, just to be straight with you dear readers, you now know I am a believer of sorts. What do I actually believe? More shall be revealed to me I'm sure. But, as you will learn, my beliefs are not so far-fetched when one carefully examines them using modern scientific discoveries. But I grant you my beliefs might at first sound anything but scientific. For example, I believe I am a simulated being having a human experience. I believe I am one with the Simulator of Our Universe. I know that we humans are already simulators and growing as types of underling super-simulators now that META, GOOGLE, TESLA, Gemini, GPT5 and others have entered the picture, but to be clear about it, I am certain that we are not The Head Simulator who created Our Simulation. On the other hand I believe that I am, along with everything else, in some way or another, one with The Head Simulator. With this admission, I continue my investigation into the mystery of Who is Us knowing that you now know my belief structure and know that it is just that, my personal, evolving belief structure. On the other-hand, as you will learn, my belief structure is backed by some solid, recent, scientific discoveries and events. It's through these discoveries and events, and the knowledge that more are coming very soon, that I think and hope that us humans my finally come together in a meeting of minds before it's too late. Okay, okay maybe we can't, maybe we will go for total global annihilation instead, but given that alternative, don't you think it might be worth a try? Whatever you think, if you read on, I think you will find that I have some very cool things to share.

Except for the above admission, much of the previous beyond-belief state-of-affairs was featured in *OSAU-2,* a first of its kind book, the literary art-form I referred to above that I've named the *AWTbook™.* As I said, and will say again because it's important, such a book highlights otherwise difficult to understand subjects with state-of-the-art, voice-activated, audio-visual references. By using the tools described in this new-age book nearly impossible-to-grasp concepts, such as those involved in the basics of quantum mechanics, can be understood without long hours of study. As proof, I became so obsessed by the potential of my idea, I dared myself to write the current book, *OSAU-3,* an *AWTbook™* that delves into some subjects for which my initial academic studies left me woefully short-listed. I was also motivated by an *AWTbook™-style,* voice-activated, *8K, ULTRA HD, Attenborough Video-Documentary* that I added to *OSAU-2's* conclusion. When I first watched that video I was suddenly transported through space and time. I could see myself manifesting with other creatures from empty-space to magically emerge into a beautiful virtual world constructed *by* illusory, mass-like, wave-particles! With such a shock to my 84-years of previous thought I had no choice. I had to forge ahead. And who knows? Maybe, just maybe, we can get some answers to the questions that arise by our pondering this perplexing, hard-to-believe Universe of which we find ourselves. And maybe, just maybe, such pondering can get us closer to answering the big question, Who *is* "Us" anyway?

Our Self-Assembling Universe-3, Who is Us?

Chapter 1 – *A New Revelation*

The following conversations take place between myself and my alter ego, Sally Reynolds. Sally is the main character in *"Sally and the Magic River"*, an Audible Book that is read to you by Rebekah Nemethy, a great narrator with an amazing, voice-shifting talent.

Audible, *"Sally and the Magic River"*, Frank H. Gaertner (4:00:01). To see examples of - - people, places and things and to hear a sample of the audible book refer to Sally's website, Sallysmagicriver.com.

Sally was 10 when her story began in 2014. In the following conversations I imagine her to be an 18-year old wannabe scientist who acts as my literary foil to challenge ideas with alternatives, questions, and sassy comments. I've since come to realize, as I will explain later, that my less dominant cerebral hemisphere is very much my Sally. I had a tragic thing happen to me when I was 10. So did Sally. My life, as a result of the tragedy, has turned out to be one filled with magic. And that's just what happens for Sally in her story called "Sally and the Magic River". Interesting, isn't it?

Sally
That's great, Frank. *OSAU-2 was* wonderful. But wait a minute! Now you say that a relatively recent, *maybe even more astonishing discovery,* was barely mentioned in *OSAU-2*?!

Frank
I'm afraid so, Sally. I mentioned it but it hadn't dawned on me until just now how truly astonishing it is. From beginning to end, the atomic builders of *all* living things have been given *coded-building-instructions by the very same builders that built the instructions!*

1

Sally

Come on now. You must be joking! That's more than hard to believe. It just sounds absurd. You're telling me that the builders themselves built the system of instructions that they now use to build all the complex things that they build?! I get that we are made *by* atoms not just that we are made *of* atoms. I even thought of an analogy that helped me think of the true significance of your insistent change in the preposition *of* to *by*. Just like one of Einstein's thought-experiments I thought, "What would I think if I saw a Lego-Set assembling itself into a house?" The house would be made *of* Legos alright. That's not surprising. But what if I saw the Lego-pieces themselves somehow coming up with the knowledge and ability to do the building of that house on their own! Cool way to think about it, isn't it? When it comes to atoms, atoms act just like a Lego Set that suddenly gains the ability to Self-Assemble anything that a Lego Set can assemble, including houses.

Frank

Oh yes! You totally got my point. I bet you can even do like I do at times when I feel and visualize myself being re-assembled every day *by* atoms! Fantastic visualization and feeling, is it not? But the previous point is not what I'm talking about now. I bet that you missed, just as I did, the incredible round-about assembly that takes place everyday as we are being assembled.

Sally

Will you give me a break! I heard you say it before but I still can't get my head around it. What the heck are you talking about?

Frank

I'm talking about the fact that the instructions and the way to encode those instructions are *self-created* by *the very same types of atoms that make them!* That bunch has been inventing, self-assembling and improving *triplet-coding-double-helical, DNA molecules* for billions of years! *And,* those DNA molecules are assembled by enzymatic-molecular-machines that are, themselves, *self-assembled by the same types of atoms* that follow the same type of DNA building-instructions, etc., etc., etc. In other words, these instructions have been delivered by molecular machines *to* all the molecular machines involved in the structure and function of all living things. If this sounds more than just little round-about crazy-complicated, it is! But a lot of progress has been made through the use of state-of-the-art particle physics, quantum-mechanics, light microscopy, crystallography, x-ray diffraction, atomic-force-microscopy, artificial intelligence, advanced chemistry, genetics and molecular genetics to name a few. Thanks to years of group effort we can now *actually see the molecular world in action* in highly accurate digital detail. Literally, *ALL* of that roundabout process comes to us through the coding and construction-skills of the builders of Our Universe.

Sally

Well, smarty pants I just found something that one-ups your roundabout. How do you think atoms came up with their coding system? Do you think they might have had some help? Have you not heard of bioelectric coding? It's one of our newest discoveries. Even though bioelectricity was discovered long ago, many claims for bioelectric effects have been discarded as quack science, and then, here comes the possibility that bioelectricity makes its own bioelectric code! So, just like the round-about in the creation of the genetic code, it has recently been pretty well nailed down that bioelectricity is actually the start of it all and is the basis of life itself! Now what do you have to say for yourself Mr. Wannabe Einstein?

YouTube "BodCast Episode 175: We are Electric with Sally Adee", Original Strength (40:52-JL).

Frank

Wow! Did you ever put one over on me! Bio-Electric Coding? I had no idea! This is going to *leave me* rethinking everything. And now that I've seen the bodcast you just found, there's no question about it. It's obvious that bioelectric coding is real. And just think. ALL of that bioelectric coding as well as DNA coding have been materialized from *virtual particles and antiparticles that* annihilate each other in the vacuum of empty space!

Sally

What?! Will you please stop! The vacuum of empty space being made capable of making anything sounds completely nuts. You're just trying all you can do to compete. Nobody I know is going to believe that empty space can create anything! I sure don't and you know I believe a lot of crazy stuff!

Frank

Relax Sally. I can prove it to you. Have you not heard of Paul Dirac, the great theoretical physicist's physicist who one-upped Einstein? Check out the video I just added to our book.

YouTube *"Tara Shears Antimatter: Why the anti-world matters"*, The Royal Institute (59:42-JL or B).

Sally

No, I've not heard of Dirac til now so I'm glad you added this YouTube reference to Tara Shears' Royal Institute lecture. It's kind'a long but now that you've given me the time to listen to it, I really like it. Maybe you're not so nuts after all. It's a great review and anyone interested in antimatter will find it interesting. She also does a wonderful job at the 12-minute mark in her lecture on Paul Dirac's role in the discovery of antimatter. And of course, I love this. I just had to say, *"Hey Siri, YouTube, Tara Shears Antimatter"* to get right to her video. But now you say that Dirac is the source of all your crazy, unbelievable talk?

Frank

Mainly, but you already know a lot of the other physicists who led to this theory on how everything comes from apparent nothing. And, if you dig a little deeper, you will find there is a lot of work still going on to meld Dirac's field-theory equation with the "theory of everything". Our understanding of the complexity of things at the *nanoscale,* as described in *OSAU-2,* was only possible thanks to the many incredible breakthroughs made by theoretical and bench-top physicists in the last 50-years.

OSAU-3, this *AWTbook™*, which I'm now writing, and that you are now helping me write, is intended to be all about life at the *macroscale. But we won't ignore the little guys. We'll* take advantage of those *nanoscale breakthroughs* to help make sense of it all.

Sally

I had no idea that that was where you were taking us with this new book of ours! But why in the blazes do you want us to call it "Who is Us?" That's just sounds stupid. It's like someone saying, "Who is I or who is me, instead of who am I!"

Frank

Come on. You know what I'm driving at. That title gets your attention, does it not? Would you rather we titled it "Who are You"? I don't think so. Anyway, I'm not surprised about your wondering where I'm going with all this talk that sounds so bizarre. I'm right there with you. I didn't truly appreciate the truth behind the crazy, round-about part of the assemblers' story until recently. But, don't worry, it gets even weirder. I just came up with something else that you might find even harder to grasp. Check out what's coming in my next chapter.

Sally

Will you please let me catch my breath! I'm still not used to this. My head's still spinning from the last proclamation where I finally understood that I've been manifested from a seething sea of virtual particles! What are you trying to lay on me now?

Chapter 2 – *Inception Begets Conception*

Frank

Don't worry, Sally. I'd never try to convince you of anything that didn't make sense. That's not my style. But I think if you don't get it now you'll get it later. You always check things out for yourself. I think that the PET scan pretty much proves that the virtual world of particles and antiparticles exists. So, at this point, there's no doubt in my mind that that virtual part of our reality is the real thing. However, I'm still perplexed by a question that's much simpler than how it's possible that anything could arise from seething fields of energy in empty space. I'm *simply* god-smacked by how DNA came to be in the first place. Clearly, primitive enzymatic molecules must have existed before DNA and, Sally, one such molecule, which we'll get to later, is catalytic RNA. But just in case you can't wait to learn about catalytic RNA, I've added a video reference to our book. It's an hour-thirty-minute, detailed lecture on the subject.

YouTube *"The emergence of RNA from Prebiotic Mixtures of Nucleotides"*, Jack Szostak, Nobel Laureate, Harvard (1:30:03-JL on a walkjogrun or save for a later B).

As for now, Sally, the macro-world beckons and I'm going to need you to keep your head and give me your full attention to help me get through what happens next.

We all might wonder how we got here. But at first *blanch* I doubt that any of us need educating on the subject. For example, superficially, you and I both know exactly how we got here. And I think we can both agree, it's pretty shocking. However, that's not the half of it. I'm about to show you the details and they are *way* more mind jogging than you've ever imagined. Just try actually seeing yourself at the time of your *inception*!

Sally

Wait a minute! I thought I was conceived. What's this *inception* nonsense?

Frank

Before you were *conceived*, you had an *inception*. Or at least that's the way I think about it.

Sally

That makes sense I guess but I don't believe it's the way most people think.

Frank

You know everything I know about *my* life, Sally, but just to be crystal clear let me bring our readers up to date. *Today is August 3, 2022.* I'm 84. I'm retired. And I know *where* I am. Gratefully, I live under a marine layer at the beach in Sunny Southern California. But *how* in the world did I get here? I'll start from the beginning. One November night in 1937 a beautiful woman named Vera and a handsome man named Herb began to thrash about. We both can imagine this sort of thing but I know some people don't like to think about their parents doing it. Anyway, let's you and I fight through any squeamish thoughts and take ourselves back in time to the incredible events that occurred in the aftermath of all the hullabaloo we know to be responsible for our creation. I warn you, this part of our story is not only hard to imagine. It's over-the-top, other-worldly and just plain ridiculous. You might be startled by that statement. So, you might ask why on Earth I would be saying such a thing.

Sally

Indeed I would. I don't see anything other worldly or ridiculous about how I came to be.

Frank

No? Just try picturing yourself as a shape-shifting superhero with bilocation ability akin to *St. Padre Pio*. Take a quick look at what I'm talking about, Sally. Just say, *"Hey Siri, YouTube, St. Padre Pio's first bilocation"*.

YouTube *"St. Padre Pio's First Bilocation"*, Fr. Mark Goring (4:27-LW).

Pretty interesting, neh? In other words, on your special day and mine, things got even stranger than those pertaining to Padre Pio. We were not only transformationalists capable of changing ourselves into new lifeforms, we were bilocationalists capable of being in two places at once! However, let us not get too carried away with all of our superhero powers. Even with all that superior talent, we were definitely not out-of- the-woods. We were each about as close to death as anyone can get without dying!

Sally

What in the world are you talking about now? I used to think you were a semi-rational person. Now you've really got me worried. None of what you've been saying makes sense.

Frank

Give me a chance Sally. Just let me show you the reality to which I refer. Did I not already tell you I thought the day of our *inception* was special and quite distinct from the day of our *conception*, but maybe I did leave off the harrowing part of that special event? So, check it out. Here is the rest of the story. On the day of our *inception* we can both say with certainty that some 300-million of our male and female relatives showed up to play. And, let's just say it, none of us were planning to play it nice. In effect we had all been invited, willing or not, to take part in a deadly, no-holds-barred, winner-take-all contest. It was to be a race for our lives, a Battle Royal where only the strongest and luckiest among us would come out alive. But wonders of wonders, and let us both pause to let this really sink in, it turns out that we were the strongest and luckiest! Moreover, of the approximately 300-million relatives who assembled with each of us that day we weren't just the lucky ones to make it to the finish line and survive, we went on to transform into fully grown human beings who are the only type of life-forms on the face of our planet who survived *inception* to be capable of talking about it! I don't know about you, Sally, but aside from all that amazing stuff, I'm most impressed by the simple fact that I have not won a race since!

Sally

I won't argue that last part. You never were much of an athlete.

Frank

In the short video I've referenced, you will see in detail what I'm talking about. I'm sure you've seen something like this before but I bet you've not gone back in time to relive the moment. As I already said, on the day of our *inception* we were all given a mission with survival odds approaching zero. And, as it turns out, even though you and I were lucky enough to survive, it was not our only accomplishment. That event was also likely to be the most harrowing, strenuous thing we would ever experience. And, apparently, the alien-like, shape-shifting transformation act we performed at the last second was also a one-off. I know, that sounds more along the lines of impossible rather than unlikely. But, obviously, somehow we did it! We transformed and triumphed over a vast crowd of our soon-to-be-dead relatives. To be blunt about it, at the moment of that transformation, by definition, we had defeated a mob of microscopic, alien-like beings!

Sally

Wait just a minute! Wasn't this supposed to be a beautiful day? I've always thought of it that way. And now you're telling me I was once a microscopic creepy thing that looked like a space alien?!

Frank

Well, have a look at the next video and then you *tell* me. But there's a lot more. We can also re-experience, our *conception* as if it were happening right now! It's the next thing, and it's the one of which you seem to be most familiar. But maybe now, for the first time, you, and everybody reading our book, can truly appreciate with me the meaning of the meeting we had with our other *half-self*. And maybe for the first time truly understand the events that took place leading up to our *complete-self*.

Sally

Okay, but you still sound nuts. And I still don't get it but I can't wait to see what you're talking about.

Chapter 3 – *Conception*

Frank

Okay, but first, I forgot. Before we have a look at that video, we have another thing that might be fun to imagine. It might be difficult but let's do our best. Let's think of ourselves as a new, entirely different life-form. Let's imagine ourselves as our other half-self being an egg-*shaped microscopic being* who resides in a *Fallopian tube*. Remember this egg-shaped life-form is the critical, most-important, *haploid-X half-self* that saves our other *haploid-self* from certain death and thereby, depending if that *half-self-tadpole* is *X* or *Y*, saves our future *female or male whole-self* from oblivion.

Sally

Oh yes, I know all about that X and Y stuff. I'm supposed to be XX and, as far as I can tell, you are XY, but I've often wondered. Maybe your absent mindedness is because you're XXY or XYY - - just kidding, *I think*. But what's this tadpole half-self business? Are you trying to make me into a half-frog?

Frank

I know *you* are kidding, but I'm not. I'm being completely serious! Just wait 'til you see the video! When our two halves meet, you will see how we instantly transform in order to be wrapped up and saved by an egg-shaped, *halfselfawaiting* that resides in a magical Fallopian-tube chamber! Apparently, somehow, even though our male father-delivered X or Y half-selves were nothing more than microscopic tadpole-like creatures, all of our relatives in the Battle Royal Contest knew our meeting up with that mystical being in a magical Fallopian-tube chamber would be our saving grace. *By the way Sally, the driving force behind that mystical meeting is an interesting subject. You might wonder about the attractive force that drives our microscopic, tadpole-shaped half to meet with our egg-shaped-half. You can find that out just by asking your phone the question.*

Sally

I found a good one that gets into the hormones involved in the process!

Frank

That's great Sally. The video you just found is also a new for me. I'm glad you came up with it. Maybe we can get into all of that later but I don't want us to get distracted. It's the video that's coming up that I really want you to see! It has everything in it that I've been talking about. So, without further delay, let's get back to the strange story of how we came to be. The next step is really one for the books. Even though the egg looks nothing like our tadpole half-self, It seems that our egg half-self knew who we were and knew just what to do to save us. Whether the egg knew or was somehow otherwise guided, that lifesaving act must have worked wonders because both our halves let go of everything to allow every atom in our weird looking, tiny-little-fused-up-bodies work together to transform into another entirely new being. This new transformative act is the one that eventually changes us into larger but still very small, curled up, even weirder looking, alien-like creatures with tails. But somehow these creatures eventually end up looking like us when we grow to be full grown, walking, talking humans. So, now, let's watch the video and see just how close we came to death's door on the day of our inception/conception and witness, be ye man or woman, *it's only thanks to our feminine, egg-shaped half-self that we exist.*

Sally

Yikes! That sounds way more scary and over-dramatic than the video I found! But what you just said about us women does make me feel a little extra special.

Frank

Well, Sally, you are special and way more extra-special than you could possibly imagine. At any rate I agree, my video might be somewhat scary and for sure it is dramatic! Even if you have previously seen videos like this, I think you are likely to be mesmerized by this one. The entire, lifesaving, life-creating adventure has been recast without audio-verbal interruption. This video has, instead, an enhanced, dramatic, *Avatar*-like, action-packed, heart pounding, film-score. At least it really got my heart pounding and helped me to feel as though I was actually reliving the entire course of events that I went through in order to experience life on our amazing planet. Maybe I'm just a silly old man but the video had a deep, emotional effect. It even got me teary eyed.

Sally

You got teary eyed? That doesn't sound like you. I've never seen you get teary eyed before. This video must really be something else! And I take it back. I don't think you're a silly old man but you're definitely a curmudgeon.

Frank

That video is really something. And for good viewing, sassy-face, I'm going to recommend everyone that they get a good set of earbuds or airpods to watch it on their mobile phones. I liked the feeling of intimacy I got from my phone's little screen and airpods as I watched unfold the dramatic beginnings of my life. Also, I'm going to let everyone know about my preference for voice-activating my iPhone with a *"Hey Siri, YouTube, Ahmed Abu-Yaman Genesis"*. This phrase also works well with the voice-control button that I have on my TV's remote. *BTW, I should let our readers know that one can also type the reference link into a smart-phone's search field but I like voice activation and this particular phrase worked well for my iPhone. Also, it should work by just using a smart-phone's home-button or by first vocalizing a "Hey Alexa" or "Hey Google", etc.* You might ask, why did I use the *"Ahmed etc."* voice command? I found that saying any part of the L'Odyssee French part of the reference made it hard for Siri to understand what I was saying. What follows in our book is the official AWTbook™ reference, from which I created the vocalized *"Ahmed Abu-Yaman Genesis"* phrase.

YouTube *"Genesis L'Odyssee de la Vie"*, FX Studio / Ahmed Abu-Yaman (14:26-LW).

Now, Sally, I suggest that you do as I do. See yourself as one of the contestants in this video. As I've already stated, be ye man, woman or beast, we mammals all started out life looking like microscopic tadpole, alien-like creatures mad-dashing our crazed way to reach our *egg-half-self*. So, hooray for us! We obviously won the race and we obviously did not die! Not only that, unless one happens to be a fraternal or maternal twin, we were the *only* victorious lifeform to have managed to complete the race. So, I'm going to tell everyone to see and live the moment for themselves and experience the fact that *we are all the sole survivors of a Mission Impossible.*

Also for those not familiar with the details of the action taking place in the video I just mentioned, I'm going to add to our book the following three videos that have narrations describing the events as they take place.

YouTube *"How sperm Meets Egg"*, Parents (2:47-LW).

YouTube *"Male Reproductive system / 3d Medical Animation"*, Dandelion Medical Animation (1:21-LW).

YouTube *"Fertilization"*, Nucleus Medical (5:43-LW).

One can also get a time-lapse look at *real-life cells* going through all of the cell-divisions needed to create an embryo.

YouTube *"See a Salamander Grow From a Single Cell in this Incredible Time-Lapse / Short Film Showcase"*, National Geographic (6:42-LW).

Sally! Wait! Don't go yet. I know you've got things to do but we're not done. With recent breakthroughs in scientific and digital technologies it's now possible to go even deeper and see how cell-division takes place at the nanoscale. Go ahead Sally! Let's do it right now. I know you have other things to do but the video is only six minutes long and you assuredly have time for that. If this video doesn't knock your socks off and take you to a new mind-blowing dimension, I'll eat an old crow ice-cold and turn over to you my precious, know-it-all badge. Let's both yell at our iPhones, "Hey Siri!" or, if you'd rather, you being a girl and all, you can just activate your phone's home- screen to sweetly say, *"YouTube, Your Body's Molecular Machines"*. What do you think Sally? YouTube *"Your Body's Molecular Machines"*, Veritasium (6:40 LW).

Sally

Sweetly, my foot! You really are getting to be an old curmudgeon! You know very well that anything you can do I can do better, and I certainly can yell at my phone if I want to. But anyone who does that probably needs help. At least you know how to pick out videos that grab one's attention. And while all of Derek Muller's videos that he posts on his Veritasium Science Channel are always awesome, this one *really* takes the cake!

Frank

Just kidding about you being sweet. But you're right. Even though I've seen this molecular machine video many times, I'm still astounded. All of this atomic magic has been happening for billions of years on our planet. And it goes on in each of us humans every minute of every day with in-comprehensibly large numbers of nanoscale players building us afresh to near perfection at in-comprehensible speed, frequency and accuracy. Even though I have more than 65-years of accumulated scientific knowledge, I am still struck dumb every time I see this video. It is a real shock to my concept of reality to realize that all of that phenomenally complex activity has been going on in me successfully my entire life. Not only that, it's been going on without my even knowing it or my paying any attention to it. An unthinking part of my brain has been involved in directing either directly or indirectly all of those molecular machines to do exactly what I want and need the entire time.

Sally

I know. It's really exciting and transformative for me too. I finally get it that those nano-gadgets are in me doing all of that complex stuff so successfully from the very beginning of my life. I think it really says a lot about *the Who in the Who is Us* question. But I thought I once heard you say we were made *by* cells. At least I know you used to think like that. You once said it's like we consist of hosts of different microscopic beings that colonize us to make our structures and our functions!

Frank

Well sure. That has always been obvious to me. But it was the atoms self-assembling the nano-gadgets which in turn make the cells, that's what eluded me. I'm glad you brought the idea up. I'd almost forgotten my old way of thinking! And I'm so glad you emphasized the "W*ho*" part because all of it says a lot about W*ho Us is*. Along the same lines I've recently been especially surprised by the molecular kinesin and dynein transport-motors. They even *look* like super-strong, nanoscale people that haul extraordinarily large and heavy things around. I find it interesting that I can either think about activating them to get me moving or I can get them to move me without thinking about it at all. Either way, it's my brain communicating to those transport motor proteins to dash back and forth as their way of getting me moving to go to the store or do all the other motive things I do. Yikes, that's hard to imagine. As I write, my conscious mind issued forth this thought and my brain has me hitting the right keys on my keyboard. And I can even do this in the dark. Just now I am closing my eyes and writing this sentence. How in the blazes is this possible? I don't believe it, yet I am right now, doing it. It's freaky incredible! Thank goodness I took typing instead of PE.

Well, at least all those many years of research done by hundreds of scientists is paying off. The chemistry experiments they did show us what's happening to the molecules involved in the process. Enzyme kinetics shows us what's happening energetically when the motor engines get fired up. X-ray diffraction and atomic-force microscopy show us what the transport enzymes look like in 3-D. Microscopy using fluorescent tags that light up the motor molecules let us see their heavy duty transport work and show us the incredible speed with which they are able to move things about. In addition, all of the recent advances in AI and digital presentation have enabled us to see them do their work at the nanoscale in highly accurate, real time, motion-pictures. Have a look. You can actually see what's going on in our cells at the molecular level. I must say that if the Ron Vale video doesn't get you chomping at the bit to learn more, nothing will.

YouTube *"How the Krebs Cycle Powers Life and Death – with Nick Lane"*, The Royal Institution (55:59-B).

YouTube *"Ron Vale 9UCSF, Hhml) 1: Molecular Motor Proteins"*, iBiology (35:25-LW).

Sally

It sounds and looks to me to be impossible. I remember you telling me that every time I move fantastic things happen in me due to nanobots. That still gives me the creeps. And I also remember you saying something about sugar molecules being converted to energy by a complex cascade of reactions carried out by enzymatic nanobots that create potential energy imbalances across semipermeable membranes using hydrogen-ions. And now you are telling me that's much like the energy imbalances that occur between the positive and negative poles of cellphone-batteries when they are fully charged with lithium-ions? Do you mean to tell me we have batteries in us created by atoms? Give me a break! I also remember you saying that those imbalances produce the potential energy that's somehow used to produce the ATP fuel needed to power the gazillions of transport molecular-motor proteins responsible for working my muscles. I know you absolutely love Nick Lane's video-lecture that explains all the complexity and I see that you have posted a reference to that video in our book, but I've not had time to see it all and have only had a chance to look at the pictures in his book called "Transformer". Talk about complicated! Atoms have had to go to the extremes of complexity to create all of the building blocks needed to assemble us into reality. But, thankfully, I can already tell that Nick Lane has put it all together in his book and video lecture in a very compelling, easy to understand, interesting way. Maybe we can come back to it later? Right now it's still too overwhelming. I need more time to read, listen and think about it all. On top of that, your having told me about the energy generation that must take place and all of the nanobot movement-magic that must go on where trillions upon trillions of atoms work together just so I can move my little finger! I'm totally dumbfounded! I'm amazed that I remember any of it because it's so complicated. But it's so hard to believe, it's hard to forget.

Frank

All of it really does seem to be impossible but now we *know.* That's exactly what happens every time we move! Vast armies of atoms get to work even when we move the tiniest muscle. Ron Vale's videos prove it by allowing us to actually see these motor proteins at work hauling their nanoscale loads of cargo. And Nick Lane's Royal Institute talk and his wonderful book, "Transformer" bring it all together to explain so much about who is *Us* and how *Us* got here.

Moreover, that kind of atom magic doesn't just stop with movement. There are other very important communications that we make every second consciously or unconsciously. And there is one such communication that is truly magical and totally unappreciated by most.

The placebo effect tells that story. That effect tells us that it is possible to either purposely think about healing one's self or unconsciously accomplish the task by having a strong belief delivered by a doctor with a sugar pill. I'm talking about the choice one has for gaining control of one's immune and repair systems by using one's brain in special ways to command the immunological and repair nanobots that one's atoms can assemble and drive into action when needed. No doubt they get to work without one's belief system involved, but it has become crystal clear that the placebo effect is real.

YouTube *"The Power of the Placebo Effect – Emma Bryce",* TED-ED (4:27-LW).

And there is another known way to accomplish the same curative "miracle". A good hypnotist with a good spell has been shown to accomplish the same thing. That effect also suggests to me that one can even hypnotize one's self into wellness with meditative processes that get the molecular machines that exist in one's immune system to get busy.

Audible *"Becoming Supernatural"*, Dr Joe Dispenza (348 pages-JL).

YouTube *"Law of Attraction and Meditation with Dr. Joe Dispenza"*, BLISS YOU (1:21:51- B).

Sally

No way that any of that can be true and yet I just saw it for myself. Dispenza showed before and after pictures of a woman's brain. The woman had frontal lobe dementia and brain scans clearly showed a good part of her frontal lobe to be dead. However, she practiced Dispenza's recommended regimen of meditation where subsequent brain scans clearly showed a return to life by the restorations of a large part of her damaged brain. Again, I didn't think anything like this would be possible but it sure looks like the real thing. I'd really like to know more about this woman. I think knowing about this woman and seeing those amazing pictures could help us all to become self-healers. But I also think it will take more than that. Why is this not headline news? Why don't more people know about it and practice this type of curing meditation? Why am I still a skeptic? I think we will need to follow up on this and prove it to ourselves.

Frank

I agree. The suggested practice of meditation is well known and certainly can be safe to do under a good doctor's care. But it takes dedication and based on the placebo effect it takes a strong belief that a meditative method will work. I've not done this before now but I've just started using some of the methodology to see if I can reverse the decline in my aging body. At age of 84 I think I've begun transforming my weakening mind, body, balance and endurance with a daily early morning routine that connects to each issue. I write, stretch, meditate, lift weights and walkjogrun four to six miles as I justlisten to enlightening YouTube videos or audible books. I'm also learning to listen to my molecular machines as they try to signal to me that something needs help by their delivery of little dull to sharp pains, signs of increasing weakness, or visible signs of physical decay. For example, I've eliminated neck pain that I've had for years. I've never had abs until now. I couldn't stand on one foot but can now do it for two minutes or more. I couldn't even do one push up but can now do 10 or more. I was feeling and looking old, etc., etc., etc. Believe it or not I've begun to change all of that. Everything but my wrinkled skin has begun to reverse a little. I especially like the fact that I have abs for the first time in my life.

Sally

I think you're losing it. Do you have any idea how nutty you sound? Do you actually think you can "talk" to your molecular machines?

Frank

I usually don't talk to them out loud but I have at times and why not? It can't hurt. I'll do it now just for kicks, *"Hey guys. Wake up! I need your help. My shoulder hurts a little!"*

Sally

OMG! You *are* losing it. I'm going to have to find some dark quiet place to think about all of the weird things you've laid on me today. Maybe you're right, but you're beginning to sound like a snake oil salesman. I must say, though, it does seem that some of us humans are not just "on" something. There are so many incredible things happening as we head into 2023 and beyond. Where's Jerry when I need him? I'd share this stuff with mom and dad but they'll just get worried and think I've gone off the deep end.

Frank

Sally left. Maybe I've taken her too far, fast and furious down my dizzy path of discovery. It even feels to me sometimes as if we're not in Kansas anymore. It's just as well Sally left. I've got me some powerful thinkin' to do and I don't want to be interrupted. I'm glad Jerry is bright and a good friend with whom she can share these thoughts. It's really hard to find anyone that understands our obsession with all these new mind-blowing discoveries. I was just going to tell her about how we are made *by* atoms that have shape-shifting talents. I'm glad I didn't. The following is really freaky but how else can one think about us Us's starting off as a single cell that can shape-shift its form and function into myriad lists of other cell types that colonize themselves into completely different organs and functions? For example, just to name a few of those types in no particular order: we have skin cells, hair cells, muscle cells, bone and teeth making cells, cells of the central nervous system, eye cells, ear cells, heart cells, liver cells, spleen cells, lung cells, pancreas cells, sex cells, fingernail-making cells, digestive cells, immune cells, etc., etc., etc. And just think, all that shape-shifting is packed into a single cell that miraculously forms when a sperm fuses with an egg which ultimately, somehow, comes up with an *Us*! Wow! And, if you missed how unerring the information is that's packed into that single fertilized cell, think about the cell-divisions that end up producing identical twins.

Chapter 4 – Twins #1

As you can tell, these impossible, yet most common events, can really get one's head swirling. But now I want to return to reality and to what could have happened at my *conception.* I'm especially curious about the possibility that I could have been a twin, either fraternal or maternal. Separate germ-cells from male-sperm are always the source of fraternal twins. So fraternal twins always result from the male fertilizing two or more separate female eggs. If two eggs are fertilized to produce resultant twins both can be xx-girls or xy-boys, or one can be an xx-girl and the other an xy-boy. Obviously, then, fraternal twins are never identical. Two spermatozoon, X or Y-chromosome germ-cells from my father would have had to be the winning contestants which went on to find, join, penetrate and fertilize two different half-self, rescuing, ovarian, X-chromosome germ-cells from my mother. And, interestingly enough, I just found out that the cell divisions responsible for the formation of all of my mother's germ cells took place in her ovaries while she was still in her own mother's womb! Maternal twins are always either identical boys or identical girls.

Oh my gosh! Another "coincidence" just happened. I decided to get a short video confirming my assertion of the remarkable fact that all of my mother's germ cells formed in her ovaries while she was still in her mother's womb and, instead, I had all my ideas about what I thought I knew about conception shattered by the following short video. The mind-blowing processes that lead up to conception have left me speechless. I know of no superlatives that I can express to match what you are about to see. And it's not like I've just been blown over by breaking news. The video was made 13 years ago! Time to wake up, Frank. You are behind times. I include that amazing video below along with a review of the fertilization process.

YouTube *"Ovulation-Nucleus Health",* Nucleus Medical Media (2:39 LW).

YouTube *"Fertilization"*, Nucleus Medical Media (5:42-LW).

Back to my thoughts on twins. And now the following thought really has me wondering. Whether any pair of twins are fraternal or maternal, if one were to ask any *one of the twins, "who are you?" Would they both not answer, "I'm me of course!" Anil Seth takes this further in his book "Being You" by using a thought-experiment trip to Mars to get across the previous point. I'll summarize Anil's thought-experiment here. Suppose you live in a future where teleportation is possible. In this future a person can get to Mars almost instantly by sending a computed identical copy of one's self to Mars with an advanced 3-D printer. At the instant your copy arrives on Mars you would then exist in two places at once. However, within seconds your selves diverge because your two selves begin to experience two very different realities. However, even a year later, even though your two selves are very changed by their experiences, both would continue to insist that they are you. That would be interesting for sure. But wait, what if your Earth-self was to be terminated at the instant you arrived on Mars? Would that still be just interesting. Would your termination on Earth be okay with the you that still lives on Mars? Maybe. But it probably would not be okay with the people who loved the "you" that was left on Earth. And it might not be okay with the you on Mars if you have been in communication with each other and thereby befriending yourself. Interesting things to think about when it comes to the self, isn't it?*

YouTube *"Anil Seth: How Your Brain Invents Your "self",* TED (23:11-JL).

Having just admitted that I'm behind times I just realized that I was about to embark on subjects for which I have little prior knowledge. And then I heard myself saying, "Well now, Frank, are you not right now writing an AWTbook^tm? Are you not right now filled with questions? You know a lot about chemistry, molecular genetics and enzymology, but didn't you kind of go to sleep when you were supposed to be studying cell-division, and are you not now anxious to learn more about all the new information that's available? Well, duh, of course you are, so, let's go, Frank! Here is another chance for you to learn something new and show everybody just how exciting it can be to write one's own AWTbook™ and resolve educational-deficits."

Chapter 5 – The Cell Cycle

I just did a JL and an LW for the following five videos. I will summarize what I've learned, and then suggest that you do the same, especially if you're really interested and have similar educational-deficits. What's cool for me about these videos is that when you see the amazing cellular movement going on, *all of you having read thus far, are already way ahead of almost everybody else on this planet!* Nobody had ANY idea how cellular-movement occurred in cell-division when I studied this stuff. It wasn't til' the 1970's that we began to get an inkling, and not until 2013 that three Nobel Prizes were handed out to the ones who discovered how our atoms came up with the way to do it. However, here in 2022 with this amazing new, high-tech world of ours all of you have already had an opportunity earlier in this book to see in detail exactly how cell-division takes place at the nanoscale in Derek Muller's *"Your Body's Molecular Machines".* And I'm guessing you have already viewed this video many times. I certainly have because it's so fantastic it's taken a lot to get it all through my thick head. So, I suggest you see it again and let the wonderful complex machinery of cell division truly sink in and assimilate how incredible it is for us to be able to see the nano-world in action!

But before I go on there is one more critical part to the cell-division equation. I don't think we have perfect ideas about how this takes place but in addition to the genetic code that carries the instructions for the building of these incredible molecular machines, as referred to previously, it has recently come to my understanding that another less-well characterized code in the form of bio-electricity plays a critical role in the assembly process of all living things by signaling just where and when our molecular machines are to be made and when those machines need to get moving! I'm still in the process of reading about this new coding system as I write but here, again, is the source of my current information. It's the BodCast interview of Sally Adee's new book "We are Electric". BTW, the Audible version is fantastic! It appears that much like Morris Code, the number of electrical-inputs in a given electrical spike can create a new layer of epigenetics that governs what makes us tick.

YouTube *"BodCast Episode 175: We are Electric with Sally Adee"*, Original Strength (40:52-JL).

And now, if you are as uneducated on this subject as I am, you should be chomping at the bit to be brought up to speed with respect to cell-division at the macroscale. If that be the case, checkout the following, and as you do you might think about the new field of bio-electric coding and just what it might have to do with it all.

YouTube *"Cell Cycle and Mitosis"*, Raghavendra Rao (6:19-LW).

YouTube *"What are Haploid and Diploid Cells"*, Nucleus Biology (4:28-LW).

YouTube *"Meiosis – Plants and animals"*, Raghavendra Rao (6:46-LW).

YouTube *"Meiosis"*, Nucleus Biology (6:46-LW).

YouTube *"The Cell cycle and its Regulation"*, Professor Dave Explains (12:39-LW).

Definition of terms used in the above Cell Cycle, Cell Division Videos:

1. Interphase – A new cell functioning in its assigned role prior to cell division.

2. G1 (gap 1) – The cell increases its mass to get ready for division.

3. S-phase – The cell Synthesizes DNA.

4. G2 (gap 2) – After DNA synthesis the cell makes proteins and grows in size, nucleoli are still present, nucleus still has a nuclear envelope, chromosomes have duplicated but still exist as chromatin.

5. Prophase – Chromatin fibers condense and coil to make chromosomes. Each chromosome contains two chromatids joined at a centromere. Mitotic spindles composed of microtubules and proteins form at the centrioles that were duplicated at interphase. They then move away from each other by lengthening their microtubules to end up at opposite poles of the cell. The Attached, specialized microtubules reach from a specialized region of the centromere of each chromosome called the kinetochore. These microtubules, in turn, interact with the polar centriole spindle fibers. This causes the chromosome to move toward the cell center, also called the cell's equatorial plate.

6. Metaphase – The spindle matures and the chromosomes align at the metaphase plate. The nuclear membrane disappears. The polar microtubule spindle fibers extend from the poles to the cell center. Chromosomes move to attach at their kinetochores to the Polar fibers from both sides of the centromeres that join each chromosomes two chromatids.

7. Anaphase – The paired chromatids separate at their centromeres to move to the opposing poles pulled by their microtubule spindle fibers. Other spindle fibers also do their pulling to lengthen the cell. Just imagine the awesome job that the nanoscale, android–like molecular machines do when they get all fired up with ATP. Once the paired chromatids are separated they become full chromosomes.

8. Telophase – The final stage of cell division. Nuclear envelopes reform around the chromosomes in each half of the dividing cell. The ribosome producing organelles called nucleoli return. The main parts of spindle apparatus fall apart. Chromosomes become loosely constructed chromatin to begin cytokinesis.

9. Cytokinesis – The process of cleaving the cell into two new cells takes place.

Chapter 6 – Twins Continues With A Thought Experiment

We were all winners in the race. The egg saved us. Initially, most of us found ourselves alone. But in the following imaginary scenario, after two, mitotic, haploid, half-selves joined forces to form a diploid cell, the subsequent cell-division resulted in a rare separation of the brand new resultant "me" into two "us's". At that instant of separation we became two, separate, indistinguishable versions of a "me"! And, at that instant, we have become maternal identical twins! Furthermore, I proffer, at least at that magic moment, both life-forms are exactly, both, "us". We are two diploid versions of our self. The two cells have just separated. Before that separation we were a single being consisting of two cells. At the separation we still are two cells. BUT NOW we exist at two separate places at the same time! And then mitotic cell division proceeds as it normally would in both of our *Us's*. What's the difference? There isn't any save for the fact that the two parts of *our Us* are separated in the identical twin scenario. One cell divides to two, two cells to four, four cells to eight, etc., etc., until we have become two separate embryos instead of one single embryo. Still there is little to distinguish us, nor is there a discernible difference even as we grow beyond the embryo stage. It really doesn't become very clear to anybody, including us, that we are actually different people. Not until we have grown to have different cognitive experiences are we even able to think about it. But having had those experiences, now we know we are different. We act, feel and maybe have begun to even look a little different. Our brains have created two very different C's, aka, conscious-states, or is it four C's? More on that mind-bending thought later, no pun intended.

But let's take this idea to an impossible extreme where our two selves remain *exactly* the same in all regards, all the way into adulthood. Even if those two clones look, act, speak and feel exactly the same, would they not experience themselves as different people? I'm sure that they would but they might begin to wonder about it. This brings me to the real substance of my idea; "The only reason I am not you is that I am me". My hypothesis holds that C=Consciousness is neither created nor destroyed. However, just like particles of matter, consciousness, at least as we experience it on our present- day planet-Earth, mostly exists in non-superimposed states. However, even though it's likely that no human being currently has their conscious state completely superimposed with another, it's possible that it could happen in the near future. I'll have more to say about that later in this book but the possibility of human-conscious-superposition brings up a challenging thought. Just *Who* is *Us* if that should happen?

Chapter 7 – *All That Thrashing About*

I hadn't planned to go in this direction but all of a sudden it occurred to me that there is something VERY strange going on here and it's true for all creatures on Earth who participate in this thing we call sex. Sex in all animals is uniformly-crazy-nuts. I am referring to the insane actions of insects, plants, mollusks, fish, birds and mammals who are universally uncompromising in their use of bizarre, ritualistic and sometimes brutal acts in order to mate. Let's just check off a few.

1. YouTube *"Queen Ant mating Season / Ant Attack"*, BBC Earth (3:12-LW).
2. YouTube "Bee Mating Ritual Caught on Camera", Nature on PBS (2:26-LW).
   And then there is:
3. YouTube "Bee Mating Frenzy Ends in Death", Nature on PBS (3:13-LW).
   I told it was crazy.
4. YouTube *"Reproductive cycle of Flower Plants / The Amazing Lives of Plants, by Dr. Larry Jensen"* Mitochondria (18:18-LW). *If you have not seen this before, prepare to have any idea you might have about plants being simple things COMPLETELY changed!*
5. *Slugs are mollusks. Based on my own experience I can vouch for the fact that they have a very unique way of sharing affection. Their sexual behavior is way beyond the bizarre. Literally, in 1969, I almost ran into one of the happy couples having at it.*

A surprising true story

My family and I had just arrived in Oak Ridge, Tennessee where I was to accept my first professional job. It was to be a combined appointment with the University of Tennessee and the Oak Ridge National Laboratory. I would be an Assistant Professor, Laboratory Scientist, Thesis Advisor and Science Instructor! Wow, was I ever excited. We were to start a new Graduate School of Biomedical Sciences and I was to have a large laboratory in a large facility that employed over 100 ORNL scientists. Originally, in WWII, the facility was part of Y-12, the area in Oak Ridge focused on the gravitational separation of U-235 from U-238. U-235 was the uranium isotope needed for the manufacture of the Atomic-Bomb. Today, that facility is called *The Biology Division of The Oak Ridge National Laboratory*. It's now dedicated to understanding all areas of biology. Earlier, soon after the war, the Biology Division was created with its focus entirely on the effect of radiation on animals, especially the human kind.

We needed a place to stay the night until we got settled. So, on the day of our arrival, we took a room offered to us in some run-down WWII housing that had recently been granted for use by our new graduate-school. Early the next morning I wandered into the kitchen and saw in the dim light of dawn a pretty, serpent-like ornament over the refrigerator. I admired it for a second but suddenly I was startled. I thought I saw it move! "What the heck?!" I can remember the hairs on the back of my neck standing on end. As I got closer, I saw something that I could not believe I was seeing. Never in my young life had I seen such a thing. Until that moment, I didn't even know such a thing existed! It was a giant slug that I now know to be the common banana variety that likes to reside in wet, humid environments.

Scientist that I am, I gently took the creature off its perch for examination. Continuing with my study, I proceeded to take the beautiful slimy thing out to the backyard. However, I don't exactly remember what happened next. I do know for sure that the moment I opened the screen door to the back porch I got the shock of my life! Hanging from an extended branch of a gum tree was a long, rope-like strand of slime that stood in silvery, sharp relief to the otherwise, mostly-dark, morning background. The silvery strand had captured dawn's early light and reflected the sun's rainbow of colors directly into my eyes. Initially the scene was overwhelmingly beautiful. However, that experience was short lived. The strand, approximately four feet long, had dangling from its end one of the most repulsive scenes I've had the privilege to witness. And the beautiful silvery strand immediately transformed itself into a gross string of snot.

The no-longer so beautiful strand had at its end TWO giant slugs en-twisted in what looked like a death grip. "OMG!" It looked like they both had disgorged themselves! I learned this just recently. What appeared to be the white interior of slug-guts was, instead, the slugs' enormous penile sex organs! Sex-organ guts joined with gobs of dripping slime as the two writhed in slow motion at the end of their snot-rope. They were doing the human version of thrashing-about in great slug-fest passion! If I'd had *an iPhone, I'd of taken a lot of pictures and maybe a movie. I'm sure it would have been a hit and one for the books on slug-slime sexology! However, as hard as it may be in 2022 to imagine, in 1969 I was unprepared. iPhones had yet to be invented. So, without pictures, until I saw the videos you are about to see, I questioned if what I'd seen, lo those many years ago, actually happened or was just a memory-figment of an over-active imagination. Thus, it is with great relief and pleasure that I am able to share the truth of my strange story. What I saw happened as I just described it. When banana slugs mate they disgorge themselves to reveal their huge sex organs along with a bunch of other stuff including slime. Please hold back your urge to gag and enjoy the following video. I wonder. Was it love at first sight?*

YouTube *"How do Slugs Mate? Explode? Mini-Documentary"*, GolovastikFilms (5:53-LW).

Hold the presses!! I never thought I'd actually see it again! I just came across another video that shows the dangling scene precisely as I remember it. Are you not impressed by the realization that turning one's self inside-out and dangling on a snot-rope is actually a normal "way" for some to do it? And are you not relieved to know that the only reason you are not-a-slug is that you are a *not-a-slug*? Somehow I find that thought comforting. If you also find it comforting, you can thank me later.

YouTube "Leopard Slugs Mating", Oldjedi (7:53-LW).

Sorry for the diversion but this is the sort of thing I expect to happen when one is writing an AWTbook™. Let's get on with the examples of odd mating behavior. This is, of course, one small but very amusing part of the story of *who isn't us but could've been.*

6. YouTube *"Puffer Fish Creates This Blue Water Art"*, Crestedduck64 (3:14-LW).
7. YouTube *"Grunion Run in Southern California"* Dana Point (3:0-LW).

Another surprising true story

I saw a grunion-run in broad daylight at noon in Baja California! (It's interesting. I could not find a daytime-event video anywhere on YouTube). I witnessed the female drilling herself into the sand to lay her eggs as males piled in on top of her in a thrashing, spiraling frenzy doing what males at orgies always do under such circumstances. BTW, I can only imagine that last part. I do not speak from experience. Also, one can only imagine the huge number of male-deposited, wannabe-grunion- haploid-halves that the males leave on the beach to die.

Chapter 8 – *Where*

So now, after all that excitement, let's just take stock and have a look at where our WHERE is. And for effect, let's affix pronouns. Let's use "I" as in WHERE exactly is it that I find myself? And for best results let's think in very-specific specifics. For example, as for me, right now, as I write this, I am a stone throw from the beach in San Diego. Yes, I guess that might sound cool for some to imagine, but maybe not as cool as where your "I" is right now, and certainly not nearly as cool as it gets when you let your "I" pretend it's suddenly transported to a lazy afternoon in October at a lakeside park in Chicago having a picnic!

YouTube *"Powers of Ten™ (1977)"*, Eames Office (9:01-LW).

YouTube "Pale Blue Dot (no Music)", Edward Crinion (3:32-LW).

YouTube "Carl Sagan Interview – Pale Blue Dot: A Vision of the Human Future in Space", Pangea (1:29:43-JL or B).

Here is a cool way I like to think about my whereness. I like to think of myself as a space-traveler who is on an intergalactic mission to the Andromeda Galaxy. My fellow travelers include some 8-billion other human astronauts. I Reside on the surface of a very tiny section of my spaceship which is protected by an atmosphere, a sphere of air through which I can freely see the night sky to witness the galaxy to which I travel as well as the edge of the galaxy that is the major part of my ginormous, spaceship-flying-machine, a machine, BTW, that includes the machinery of a large planetary system that orbits my very own life-support's energy source, a star.

(Yes, check it out if you haven't. You can actually see with your naked eyes on a light-less, moon-less, dark night, the stars in our own Galaxy, "The Milky Way"! And, also, actually see another Galaxy, "Andromeda", which is about 2-million light-years away!).

YouTube *"A Journey Around the Milky Way"*, Kosmo (1:36:21-B).

YouTube *"A Journey to the Andromeda Galaxy [4 K]"*, SEA (36:19-LW).

All of this describes the magnificent spaceship, on which we are all traveling as it zooms its way to meet-up with the Andromeda Galaxy at an incredible speed of 300,000 mph! I think we can all agree, we are on one heck of a ride on one heck of a spaceship. But please know, we are on a dangerous journey in Our Universe for which there is a very good chance that none of us alive today will ever get out alive long enough to be able to arrive at any of our possible galactic destinations. Unless, that is, Ray Kurzweil is right. He talks about AI and other technologies headed our way. They may be able to help us figure out methods for some of us to live long enough to get there, wherever the "there" of off-planet might be. It might sound like Kurzweil and others are kidding but they're not. We are on the cusp of some very hard to swallow stuff. I'm almost 85 and I wouldn't mind living a few more years beyond 100 but that would not be nearly long enough. Andromeda is over 2-million light years away! Even approaching light-speed travel it would take nearly 2-million years for my identical-self on Earth to know that the other you-of-me who doesn't age, while its traveling at the speed of light, met up with that galaxy. My space traveling you-of-me arrived at Andromeda in less than an instant! But here's the thing, If my self on Earth were able to maintain its existence in some kind of hibernating suspended-animation, that self could go to sleep to wake up 2-million years later and not even be aware that any time had passed. Everything else on Earth but the hibernating me would have aged by over 2-million years! Mind mindbogglingly interesting, isn't it?

YouTube *"The Milky Way and Andromeda Galaxy Collision Has ALREADY Begun!"*, Voyager (10:09-LW).

And now that we all know where our where is, let's see what Sally has to say about it.

Chapter 9A. - Movement and an Example of How to Write an AWTbook

Frank

Hey Sally! Are you there?

Sally

Present and accounted for. I took a nap but it didn't clear anything up. However, I am ready for more of your nonsense. I just saw the wonderful Eames video and the colliding galaxies.

Frank

So you know I'm not just blowing smoke. I'm just taking us where the questions asked take us. Now that we know more about where and what we are, let's compare notes. I'll start, if you don't mind. First I get that I am like one of the *"Who"* in Horton's story.

YouTube *"Dr. Seuss" Horton Hears a Who!"*, Preview (2:01-LW).

I live on a speck that we call Earth with a larger hierarchy of other specks that include 1. a star we call the sun; 2. a still larger speck we call the solar-system; 3. an even larger than that speck that we call The Milky Way Galaxy; 4. a ginormous speck we call Our Universe and last but not least; 5. a speck that dwarfs them all, if it exists, a Multiverse within a Void. But let us not forget why even the multiverse might be regarded as a speck. If the multiverse is compared to an infinity that contains it, it's a speck in the *void*. So that's *where our where is*, but we're not done. Heading in the other direction to the tiniest of scale, we find ourselves to be the dominant speck in a descending hierarchy which includes the macro-visible world of tissues down to the micro-visible world of cells to the nano-pico-femto-visible world of molecules, atoms, and quarks all the way to the ethereal-not-so-visible world of wave-particles that manifest themselves from virtual particles-and-antiparticles that flash in and out of existence from fields of empty space.

How does knowing all of that make you feel, Sally? We are on a planet with a mass so great that it affects the curvature of space-time's hold on our silly selves such that we stay stuck to the surface of a blue ball that's comparatively so small it rapidly becomes almost invisible when viewed from the rings of Saturn. Nevertheless, that teeny tiny ball of ours supposedly warps space-time enough to give us the thing we call gravity.

Sally

How does that make me feel? I can't tell you that. It's way too much to feel anything but confused. Other than that, I can tell you that you haven't made me feel any less ephemeral! Not only that, as long as we're handing out crazy stuff, I'll try to add to your handout and make things even crazier. We are also held in rotation about a star by the same space-warping effect brought about by our star's even larger mass. And, if that's not enough, we rotate with nearly half-a-trillion other stars about a supermassive black hole. In one fell swoop we've not only been enlightened by the Eames' video depiction of where our "where" is, but also we get seamlessly introduced to what our virtual "what" is down to the vanishing scale of wave-particles. So, not only do we get *astronomical* views that let us confront our impossible-to-believe *"where"*, we get *nano-nomical* views that let us see for ourselves that the fact of our very own existence represents an it-can't-possibly-be-true *"what?!"*. I suppose that's great to be so confrontationally introduced to our reality but where does that leave us with our big question? What have we learned about the *who of it all?* If anything, knowing about all of the above makes the *who* question even more puzzling, do you not agree?

Frank

Yes, I totally agree. So let's review. We are human beings, the product of more than four-billion years of ceaseless reinventing carried out by a team of the world's most skillful, brilliant inventors; Hydrogen, Carbon, Nitrogen, Oxygen, Phosphorous, Calcium and Iron to name a few. Their work has been remarkable and their achievements have never been outdone. For example, after billions of years of trial-and-error, hypothesis-testing-research-and-development, this amazing, hardworking team of atomic miracle workers came up with everything we need to survive, as well as for me to be able to write this book, and for you to be able to read it! Not only that, our planet literally went through hell, high water and back to make it possible for us to be here. And if it wasn't for that asteroid we might be here talking to each other looking more like dinosaurs.

Come to think of it, having just had that incredible what-if thought, I think it would have been just a little more than cool to be able to fly to work on my own wings like a pterodactyl! And that's to say nothing about what it would have been like to be a super-intelligent *Tyrannosaurus rex.*

YouTube *"Mankind Rising – Where do Humans come From",* Naked Science (43:32-JL or B).

YouTube *"The History of Earth's Five mass Extinction Events (4K)",* Spark (47:55-JL or B).

And check this out Sally. A video I just now added to our book is a nanoscale example of the atoms in action demonstrating one of the atoms' most amazing inventions. It's the roundabout to which I referred at the very beginning of this book when I first amazed you with how our atoms invent and self-assemble everything.

YouTube *"DNA Animation (2002-2004) by Drew Berry and Etsuko Uno",* Wehi.tv WEHimovies (7:19-LW).

Sally

That video *is* wonderful! But why did you add it here at this point in your book? It seems out of place because I heard you mumbling. Weren't you just about to start sharing your ideas about God or at least some form of Game Master, a GM-type of Supreme Being who creates Our Universe? Doesn't that fly in the face of your idea about our being in a simulation where the nano-forcefields we call atoms invent and create everything on their own? Not only that a lot of people who believe in God don't believe in evolution much less a simulation and even physicists cringe when you anthropomorphize atoms by talking about them as if they were alive. What about that? How are you going to fit all that together without ticking *everybody* off? How are you going to fit evolution into a Creator God or Creator Superior Being of some kind? And why do you think you can get away with talking about atoms like they were nanoscale human beings. Moreover, if you want to know what I think, since we both know you don't know what you are talking about, I think you should run just as fast as you can for some neutral ground before you get tarred and feathered. At least why don't you refer to your idea of an *Evolving-Wannabe-God-Some-Day* or your idea of a *Personal Supreme-Being* as a hypothetical SB for short. God's *real* for me but you are so half-assed about it you had better hedge your bets.

Frank

Okay, okay already! You're right. I do get carried away. So I think your SB idea as reference to something hypothetical for the anti-religious to ponder might be perfect, especially since my idea is probably BS. Obviously, we can't know the mind of a Deity, but I think we can get into the mind of the inferior, primitive-stage SBs that we are about to become. Why do I say that? Just as we are beginning to reveal in our discourse, Sally, the makings of a hierarchical SB already exists within the growing multitude of global Game Masters, aka GMs. These GMs already operate in their little, virtual-reality, role-playing, computer-game simulations. We'll get into the evidence that we are indeed in some kind of Grand-Simulation later. But even if we are in such a simulation, it is a simulation where atoms do their creative building things on their own by their individual, electromagnetic, chemical, building-block algorithms that manifest everything. The existence of a hierarchy of GMs begetting SBs changes nothing about our Self-Assembling Atoms and their role as the Assemblers of Our Universe. And, as for my anthropomorphic description of atoms creating molecules that look a little like nanoscale versions of human-like creatures, *I* sure didn't do that. I showed you in the Molecular-Machine Video that the atoms themselves made the molecules look like little humans. So, I've already shown and explained my anthropomorphic use of the word "BY" in great detail. However, if you didn't get it the first time around, Sally, try this on for size. Atoms *create*, or if that word *create* has you tearing your hair, atoms *form* molecules. They are able to do that by way of chaos theory and the force of Free-Energy, i.e., the energy available to do work. This energy exists according to the algorithms governing entropy and the entire simulation of Our current mathematical Universe. Molecular creation, or if you prefer, molecular formation, is the algorithmic-means by which atoms will eventually grind everything down to the lowest possible energy-state. That is, according to the laws of thermodynamics, the energy to do work will ultimately approach exhaustion to be replaced by nothing but weakly giggling, useless atoms. Entropy, i.e., the part of the thermodynamic algorithm that drives everything, will theoretically drive everything to a stand-still. All that will be left is an ice-bound universe at the end of it all. In other words the activity of all useful work will eventually dissipate to approach zero, which as a consequence, will essentially bring everything to an effective, withering halt, rather than a screeching, frozen stop many trillions of years from now.

By the way, I just had another of my not so brilliant ideas that I just must share. Arguably, we are the current supreme beings on our planet, but our status as such is rapidly increasing. We may all soon be able to neurologically link to supercomputers by way of NeuraLink or some other such device. If all humans on Earth were to be so linked our status as an SB would suddenly escalate. Now, here comes the interesting part, once linked to such a supercomputer's AI, that AI could operate as a single source SB for all of us. We would all in essence become that SB. Our bodies would need to be managed by robots because in that mode our Personal Responsive Conscious State would cease to exist. This would be similar to the way such happens when we sleep and experience our daytime-self leaving us. We seem to blank out in NREM and then start dreaming ourselves into an entirely different state of reality in REM.

Sleepfoundation.org, *"4 Stages of Sleep: NREM, REM, and the Sleep Cycle"* Kendra Cherry (Feb 2023).

By becoming that SB we would in essence all become a single SB. That SB would likely be silicon-based, and no longer a carbon-based being. That SB could travel to the stars at near light-speed. Gravitational effects, atmospheric effects, emotional effects, etc., would not impact that SB. Any human could detach from that SB at any desired time, but just as is true with those who play with computer-game virtual realities, I'm guessing many would find no need to detach from their "real-life" Star-Trek adventure as that SB arrived on a new planet. Gravity, Temperature, Atmosphere and Lighting effects do not impact your SB. You are on the adventure of a lifetime. But let's say you die in the meantime. It does not matter. You are one with the SB. If you want to continue life as a carbon-based human being, even though the original you on Earth may be dead, you would be able to return to humanity and be any person that you wanted to be.

Sally
Oh my! Oh my!! Good grief!!! I knew you were going to do this. Even so, I can't say I was ready for it. But I don't plan to be misled. You really don't know what you're

talking about, do you?. However, I will grant you, there might be a kernel of truth to your rantings so I'm going to stick around to see what happens next.

Frank

I admit that was a bit of off-the-charts Science Fiction. But maybe this will bring you around to my way of thinking. I just found out today that I'm not the only one with such challenging controversial thoughts. Check out the following amazing interview in which Richard Dawkins gets himself wondrously interrogated by my favorite interrogator, Mr. Lex Fridman.

YouTube *"Richard Dawkins: Evolution, Intelligence, Simulation, and Memes"*, Lex Fridman (1:7:20-JL).

Sally

Whew, that's a long interview. But thank you! It was so interesting and so intense I couldn't stop listening on my run this morning. And you are definitely not alone. Your out-of-this-world thoughts match up with some pretty incredible company. Nonetheless, I still think you're bats!

Frank

Thank you. As you know I take great pride in that bat-fact. But enough with the speculation, lets get back to reality. It's clear that the designing activity of an atom is not that of Intelligent Design. Nevertheless, I think you will agree, the Self-Assembly of Our Universe *by* the so-called inanimate world of atoms is still a fantastic achievement of design. How can that be? And does that mean there is no Intelligent Design? Maybe Dawkins says that Intelligent Design doesn't exist, but I don't say that. Our Universe is mathematically based and if there is an algorithmic simulation behind it, there must be a *Simulator*. However, If there is a Simulator with the Intelligent Design of a Universe in mind, how does the Simulator go about it? If the concept of a hierarchy of GMs begetting SBs is correct, all of those SBs would work with arithmetic algorithms that would have them, at all levels, creating games in which they could be involved in full-participation, much as the gamesters of our present and near future might find themselves fully participating with AI-assisted avatars and brain-computer-interface links. And here is what I think is obvious and very important to understand. There is plenty of evidence that no SB would ever want to know what happens next! Why so? Just think about it. There would be no entertainment value in knowing what happens next. (PS, Also, unlike Dawkins, you now know my idea about the true meaning of life. Forever is a long time, and because of that anything that *THE Ultimate Supreme Being* did would of necessity need to be a VERY HUGE form of *foreverlasting-entertainment* or, if one prefers, *experience*.) Our SB might not be the Supreme One but *is* likely already one who lives forever, or at least is well on its way to doing the business needed to do that. To come up with such a profound game-set, as would be needed to shape Our Universe, the wherewithal of the Simulator would need to be highly advanced. How advanced? Maybe infinitely advanced, meaning not definable. In any case, no SB would waste time whiling away infinity by pushing atoms around or by any other means communicate with each atom in the Universe to tell each atom what to do. Nor would a Universe-creating SB likely even be interested in orchestrating just how the game plays itself out. However, having created such a game, isn't it obvious our SB would want to be intimately involved in it? Our SB could time-share with *Us*, that is the part of *Us* that resides within the Simulation of Our Simulator's Universe. We could be like AI-assisted avatars, or in some other way fully participate in the simulators game. Obviously, it's however Our SB would want it to be. Am I demeaning God? Not at all, I'm just searching for a way to visualize for my poor, little self how the fantastic show in which we find ourselves came to be.

To beat this concept into the ground, Sally, in case you didn't quite get my drift the first time around, I'm deliberately trying to make an analogy with respect to something a current-day gamer might do in a role-playing game equipped with a brain-computer interface wherein the game's avatars are themselves equipped with AI. Even the devoutly religious say that we have been made in God's image. So it doesn't seem to me to be so outlandish to think we might try to make a parallel meeting of the minds here. Anyway, let's say I'm right and something like this is going on with *Us* where we are our SB's avatars equipped with our real-life form of consciousness. If so, being such an avatar, would not one want to reverently communicate with one's GM, aka SB, aka God? Would not, then, praying make a lot of sense? I think the only questions might be, how interesting an Avatar one might be and how good one might be at communicating in order to make a good connection with one's SB. It might also matter as to how ancient and how advanced our GM, aka SB, aka God might be? Would a 1000 years be enough time to create the algorithms that govern Our Universe? I don't think so. Maybe, even, multiple trillions of years would fall short. Maybe our SB is far beyond it all and is no longer a galactic being and may not even be a part of anything that even resembles Our Universe. Maybe our SB is the Personal God that many on Earth seek for solace. Or, maybe, our SB is indeed The Master of the Void and the One who created The Simulation of Something from Nothing that we know to be Our Universe. On the other hand, maybe Our Simulator is a relative Junior GM who is low in a multiple zillion-year hierarchy of GMs in a hierarchy of multiverses. So, where do we stand? I think there is almost no question that we are living in a simulation. Given that to be the case, the next question is, what is the nature of Our Simulator? Since I think one can really only know that through faith, it might not hurt to sing in a choir. Aside from that, what have we accomplished here? I think it turns out that there is good reason to believe that we conscious beings are an important, integral part of the Simulator's Simulation. I look around. All I can do is say, "Wow! Well done! This is Astounding, Thrilling, Wondrous, Amazing, Fantastic and Startling!. Thank you, Gah, for the experience."

Obviously, except through faith, there is in no way anyone can actually know these things but I find the significant clues from science and the blessings I receive from the magnificent, beautiful display our starry Universe gifts us each and every day gives me hope that we humans with our diverse ideas and beliefs can join hands in mankind's incessant search for answers. Whatever one knows or believes, whether it be from scientific discovery or simple faith, can't we all agree, at least, that there is something impossibly amazing and wondrous going on in us and all about us? I can't help but applaud the Author for the Earth-Rises and Earth-Sets, Rainbows, Butterflies, Humming Birds, the Beating of my Heart and the Love from people and other animals that grace my existence each and every day. But, oh my. It gets even better. Is it not also entertaining to think that we might vaporize ourselves as we almost did in the 1960's, or be eradicated like the dinosaurs were by a large meteor, or wiped out by methane belching forth from tundra, or something else bubbling forth from the deep sea, etc.? BTW, I and my family are especially blessed, at least as far as the end of the world goes. Since we live like we do near a major military base right here in San Diego, we could witness the world as we know it vanish in a flash! I only add this humorous downer to point out that for any entertainment to be good entertainment, there must always be upsides and downsides. Does this not answer the question many ask, "If God exists, how could God let such a horrible thing happen?" I say God does not know or want to know what happens next, and if it was always sugar and spice and everything nice, nobody, not even God, would be able to stave off boredom.

Sally

Yikes!

Frank

I know this is a bit too much so let's get back to the argument over the reality of evolution. It makes sense to me that the "Master Theory of Everything Algorithms" are also the key to understanding evolution and what atoms do. Everything is ruled by Big Bang, chaos theory and thermodynamic algorithms that lead to atoms and their dissipative-structure, fitness-algorithms, all or which govern the evolution of Molecular Machines and DNA Genetic-Code Roundabouts, and lead to electrodynamic and bio-electric cellular-communications that ultimately come up with living things. Only the fittest, most-likely-to-survive, molecular machines make it to the next level for a chance to arrive at that act of survival. Atoms build molecules by virtue of the Pauli-Exclusion rules of their electron clouds and rules of thermodynamics demanding that lower states of energy be met. And all of that will continue until Our Universe grinds to its maximal entropy-induced approach to a stop. For example of the Pauli-Exclusion Principle in action throw some metallic Na-23 into water and watch it make sodium hydroxide and fire! There's lots of YouTube videos on this wonderful experiment. So be my guest. I'll let you call up Siri or whoever you have on your phone to check it out.

wikipedia "Pauli exclusion Principle" (one page). web.mit.edu "Complex Adaptive Systems-MIT" (9 pages).

Just behold all of the successful lifeforms that have made it to enjoy this process, at least for a little while. We are but one of the many successful lifeforms that find themselves *vastly outnumbered* on this planet. Just think of all the microbes that lurk on and in our lumbering hulks. However, Sally, enough of this mind-stretching harangue, at least for now. As to your other point about writing, that's exactly what I wanted to show by adding the above discovery ramble into this abrupt, seemingly out of place spot of our book. I wanted everybody to see, including you, Sally, just what I think is so wonderful about writing one's own AWTbook™. It will happen that one will make exciting discoveries as one queries the AI on one's phone while one writes. Where one adds the newly obtained information is entirely up to the author. But, once added, that exciting discovery has been archived and now exists in the reference list of one's very own book, a process that fixes the new information in one's mind and makes it available for future viewlistening by you and others. It's a very powerful learning tool and a fantastic way to rapidly build one's knowledge, especially when it comes to one's career-development. For example I just found the DNA and Lex Fridman videos to be extremely interesting and wanted to keep them handy for myself by adding them to our book. And I'm also excited to be able to share them with everyone else who reads our book. The videos add a lot to one's understanding of the miracles behind *what* we are and how we came to be. All we had to do was make a place in our book that we think is a good place to add them. That's all anyone has to do. And now that the references exist in our book, we can go back to them whenever we want. Moreover, each video that one adds will lead to others of great interest. I guaranty it. For example the next two videos just showed up as great examples of how our n*anoscale inventors made it possible for all of us humans to rise to the level of* conscious beings, beings who are, I'm sure, the only such beings on this planet aware of their finite existence as we orbit a star in a half-trillion-star galaxy. And, let us not forget, our galaxy exists in an expanding universe consisting of multi-billions, maybe even trillions, of other such galaxies.

Sally

Well you've just confirmed it. You are somewhat beyond being a crazy person. But a lovable one and I have to admit, did we all not get a little closer to understanding *Who is Us*?

Frank

Indeed, I sure think so, and there will be a lot more to say about *C=Consciousness*, but you will find the following videos I'm adding to our book to be useful teasers.

YouTube *"Neuroanatomy S1E1: Introduction to the Central Nervous System"*, UBC Medicine-Educational Media (14:47-LW).

YouTube *"The nervous system in 9 Minutes"*, CTE Skills.com (9:22-LW).

So, Sally, here we go. As I've already said. You and I are writing this book by adding videos and other web references as we go. And it's exactly what I propose everyone else do when they write their own such book. Have an idea, ask a question that goes with the drift of one's ideas, follow the video or videos one's phone-*AI* produces etc., etc., and manually write or just talk to your phone and have it do the writing for you. I am finding that new questions and ideas come up as if by magic. For example as I was just writing my part in this book, the previous nervous system video just led me to a profound realization.

Sally

I get it but what's this? Another one of your off-the-wall revelations?

Frank

I'm still thinking about my fingers. I can't get over how they work. Take a look at your fingers, Sally. Now wiggle them, and then think about it as you say to yourself, "How in the heck did I just do that? Ask yourself, what is it that allows me to do such a thing? How is it possible for me to move so specifically without even thinking about it? It's even so that if one thinks about moving, one's movement gets much slower. Just try wiggling your finger halfway. I, at least cannot do that fast initially, but I can do it fast once I've thought about it and stop thinking. Now you can be impressed with wonder as the pianist in the next video appears to do the impossible, and do it better than anybody else on this planet in my opinion. There are many fine pianists that perform this piece with great skill. But for me and many of her followers, Polina is the very best. However, I have not chosen to show you this video, Sally, just because of the speed of Polina's finger work, which is indeed unbelievable. I chose this one to show you because I am obsessed by this work by Bach and by Polina's performance of it. And I simply want the whole world to experience what I see and hear in this video. It's nothing but uplifting and wonderful to see and hear what this particular one of *Us* can do. This YouTube video is by far the one to which I've watched and walkjogrunlistened to most often. But it's just one of my very long, treasured list of such musical videos. I so very much hope this one rings your chimes as well.

YouTube *"J.S. Bach Concerto No. 1 in D Minor BWV 1052 Polina Osetinskeya Anton Gakkei",* Maxim Novikov (23:52-JL or B).

Sally

Well, duh. You sound like I've never heard it before. Of course it rings my chimes! Although, I don't think it made a lot of sense to add here in our book. Nonetheless, I'm really glad you did. I, too, am in awe of the way Polina plays. I found one thing very interesting though. I've never before seen her play from a musical score. All the other times I've seen her play she's had the music memorized. But, in this Bach Concerto we see her intently reading the music and using the help of a very serious-looking page-turner. Strange. I wonder why? Maybe its because this particular work and her performance being so special was no accident. It's clear that she played with GREAT intention and did not want to miss a beat. There is no question that It IS special. In my opinion her performance of this work by Bach is unmatched. I think I can safely say, nobody is going to outdo this one. I can say that because even though it's "stuffy old" Bach, *it is one of the most popular videos on YouTube.* And for this discussion, there is no question, she can really move her fingers! But, as long as we are at it, I have a surprise for you, Frank. Take a look at the following video. It's a fantastic Concerto for four violins by Vivaldi that's been orchestrated by Bach for four pianos. This one is also a favorite of mine, and of course it goes without saying, yours.

YouTube   *"Bach Concerto for 4 Pianos. Multipiano/Tel-Aviv Solosists/Barak TI"* Soloists1 (10:45-JL or B).

Frank

Thank you, Sally! This Vivaldi 4-violin concerto is incredible, and it so incredible Bach liked it enough to orchestrate the work for four harpsichords, played here by four incredible pianists. But I'll just say to everybody, don't miss the string-version. The finger work on violins is totally different from that when played by four keyboards. To tell the truth, I love both versions, Bach's and Vivaldi's. Bach must have been just as taken by Vivaldi's work as the rest of us so he couldn't resist hearing what an amazingly different kind of musical experience could show up when four well-tempered claviers played it instead of four violins. In fact I like both versions so much I've added both to my downloaded walkjogrun music-list. But, boy, are these two videos ever fun to watch! To see the finger work of four violins versus four pianists is a kick! However, let's move on. We're sort of off the topic.

Sally

You think?

Frank

Alright already. Now it's time for you to give *me* a break. I, Frank, am the one who thought of this book. So, even though you are helping me write it, I get to do what ever I want with it. The following expose's should shed some light on the mysterious subjects of intentional movement, and the ability to do it so specifically with such great precision without even thinking about it. But, be prepared. We are going to plunge even deeper into the world of science-fiction becoming science-fact, as there will be a lot more to say about the way brains operate when we explore AI and robots.

YouTube "New Brain Implant Begins Human Trials-Neuralink", The Tesla Space (14:01-LW).

Sally

Okay, "I Frank, the one who thinks they're so hot". You might of thought of it, but you would be hopelessly lost, and I'm pretty sure, fresh out of luck when it comes to typing this thing. I've saved your silly tail-end many times already. So, you're welcome. And, come to think of it, I think without me you'd probably be long dead by now and co-dependent that I am I'd probably would have tried to save you and ended up going with you on that trip. Helping you, indeed. How conceited can a person get?!

Frank

I'm Sorry, Sally. I forgot to whom I was talking. Please forgive me.

Sally

I'll think about it.

Chapter 9B – How Insects Move

Frank

And now, Sally, now that we've seen a little about how humans move, just for fun observe flying insects doing the amazing things they do with their ever-so tiny brains. They, definitely, can't be doing a whole lot of deep thinking about moving anymore than we do. So, *be* the insect and marvel at how agile, how precise and how quick *you, being the insect*, must be when you're not in slow motion. Ask again, as you are being the insect. Think *"How in the world did I do that?"* Your immediate intention and response with such a tiny brain seems impossible, does it not?

YouTube *"Praying Mantis and Move! 15 Insects Flying in Slow Motion"*, Ant Lab (8:23-LW).

Sally

I must say. That was one of the coolest things I've ever seen!

Frank

You ain't seen nothin' yet. The house fly with its ability to avoid swatters maybe one of the world's most underrated, underappreciated, aerobatic performers. No matter how annoying flies may be, there is no denying that their flying ability is fabulous. But how do flies do what they do? I didn't know until I saw the following video. Check it out. House flies and other flies like it have a cool device called a haltere. This neat little attachment is key to knowing how a fly does its amazing stunts. But as remarkable as it may seem we sometimes forget that all of us creatures have been given mind-boggling special molecular devices to give us the coping talents we need. And don't forget, all of these inventions are nothing but the pure genus of an *inanimate world* of atoms that accomplish VERY patient hypothesis-testing experiments taking many billions of years to complete as they take hints from failed experiments for what they must do next to come up with a particular gadget to enhance a creature's fitness.

YouTube *"Here's How that Annoying Fly Dodges Your Swatter"*, Deep Look (4:34-LW).

But wait, you still ain't seen nothin' yet! You've already seen Kinesin and Dynein protein motors in action during the earlier cell-division video, but this short one on Kinesin is the cat's meow. It shows how this engenus molecule easily hauls what looks like a massive boulder down a microtubule fiber. These motor protein molecules apparently have no brain, *apparently,* but they sure do act like they have one. They move and act like they know precisely where they're going, what they're doing, and when they're going to be doing it. Just like us, and maybe even in ways better than some of the macro-us, these nanobots obviously have the ability to move with very precise intention.

YouTube *"Kinesin protein walking on microtubule"*, Emanuel Dumont (0:22-LW).

YouTube *"Drew Berry: Animations of Unseeable Biology"*, TED (9:08-LW).

Sally

I thought that was a joke the first time I saw the kinesin molecular motor doing its thing. And it has to respond instantly at high speed when we either move without thinking about it or move with intention by actively thinking about it. How is that even possible?

Frank

I don't know but it's got to have something to do with the electrodynamics brought about by electromagnetism. Maybe if we just observe Kinesin again and question the molecule as we watch it move. Maybe that will give us an "aha". How does that motorized molecular transport-machine know what the heck its doing and when to do it? You can think that ability to *know* is just built into its structure by the laws of nature or the Game Master's algorithm. But that tells us nothing, does it? I submit that we creatures have a hand in it by our very existence and we human creatures have it over all other earthly creatures because we have a level of consciousness that lets us think and talk about it. All creatures have autonomous parts to their nervous systems. The autominous part of our central nervous system runs on automatic. We don't *have to* think about it. We have another part of our central nervous system that lets us *will* our bodies into action, and we don't have to think about that, but we also can chose to do so! I just typed this without thinking about it. My fingers move but what I am thinking about is not the movement of my fingers. I am thinking about what I write. But I *can think* about the movement of my fingers. I will do that right now. Okay index finger, I am going to think about you, specifically, typing the letter "z". Alright Frank, stop right now for a second and try thinking about typing that letter and then do it five times, zzzzz, OMG! Was that ever difficult! I actually had to force myself to *think hard* about doing that. I can't get over how interesting this is.

Sally

You're just being silly. I need to take another walk and clear my head. I'll see you later.

Frank

Okay, don't go far. I'm really going to need you. I can feel myself leaping off into the deep end. So, please, before you leave have a look at the videos I'm now adding to our book.

YouTube *"The Inner Life of the Cell Animation",* XVIVO Scientific Animation (3:13-LW).

You've already seen kinesin in action but in the above video you get to see kinesin at work in the context of its cellular environment.

YouTube *"Dynein Motor Protein"*, Molecular Animations of the Cell (3:00-W).

Sally

Thanks. That *was* so cool. But I really do need to take a break to build up my reserves now that I know what comes next.

Frank - See you soon.

Chapter 10 – Consciousness and the WHO of the Matter

Thus, there is no question, even at the nanoscale intentional movement can be seen with atoms and molecules. And highly elaborate movement can be seen with all animals, insects and plants. That's great but when even the nanoscale can do it, it doesn't seem that intentional movement says much about Who is *Us*.

If not movement, then, what does have something more to say? Let's get right to the chase and maybe, then, walk it backwards. I am a human being. I call myself Frank. I can see, hear, read, write and think, I think. What, then, really speaks of WHO "I" is. Well, I guess it's obvious. The thing is, there is no see, hear, feel, smell, read, write or think without C=Consciousness. It's also clear that responsiveness can, indeed, be a way to identify that a being is conscious. It is also obvious that there are different levels of C. A fly can surely see, hear, feel, smell and respond in a number of ways but it also, *surely*, cannot read, write or have deep thoughts. However, I think a fly must have some ability to think, i.e., process information and act on it. But the way flies do it must be considerably more primitive than my full range of ability when it comes to information processing. Okay, so I am clearly not a fly. I didn't need to know anything about how a fly thinks to know that much about myself but thinking about the fly thinking, versus my way of thinking, is probably important to our understanding the nature of consciousness, especially when one gets down to understanding that *Who* of the matter. I have another thought experiment that might help us ponder this subject a little deeper.

Chapter 11 – Deep Thoughts

A lot popped into my head this morning as I was waking up. I don't recommend this, but I think I need to share it. It's a practice I've done for over 30 years. The practice evolved from problems I was having. It turned out that the problems were due to the fact that I am very sensitive to caffeine. Too much caffeine can give me severe symptoms of stiff neck, lockjaw, lump in the throat, neck pain and headache. I first tried to quit drinking coffee by substituting it with tea which solved the problem *for a little while*. But, as it usually goes with addicts, escalations can accidentally happen. And I, no kidding, started drinking caffeinated sodas by accident. Did you know that orange soda contains caffeine? I didn't. All I knew at first was that the orange soda I had swiped from the office frig at work when I was desperately thirsty worked like a miracle. After seven years of little or no caffeine it was like I had just woken up. I thought, "It must be the vitamin C". Then I looked at the label, no C but plenty of caffeine. I was again hooked. This state of addiction could not stand so it led me to a *bright idea*. "I'll just treat caffeine like the powerful drug it is!" I began using a nonprescription, 100-mg, pill-form of caffeine and precisely dosed myself with what is roughly equivalent to one cup-of-coffee three or four times a day. Since then, I've been able to use the following routine without much variation for the last 30 years. I usually hit the sack between 9 and 10 pm and wake up between 5 and 6 am. The first thing I do is take a half-pill and go back to lying down while I'm still in a semi-dream state. It's then that my imagination takes off and the ideas start flowing. For example, the following *deep thoughts* came to me this morning as I was thinking about an earlier passage in this book. As you can tell, I Frank, am really struggling with this. I think a REALLY BIG SHOE is about to begin, and I want to get as ready for it as possible by writing about it. Please try to be patient with me.

Forever is a long time. I feel like I somehow showed up in the middle of forever. But how is that possible? I don't think it is. Infinite time is undefined. I know just one thing for certain, and it is the ONLY thing I know for certain. I am a conscious being and in some form or another I exist. So, that means consciousness either just showed up in me somehow and I have somehow showed up in the middle of "forever" or consciousness, of which I am undeniably a part has always existed or had a start and will now always exist into infinity. Herein, my C stands for Consciousness and in this model it is just like Energy, neither created nor destroyed. And also, just like the E in E=mc², C can be transformed to I=Information where I is nothing more than one of the transformations of E. Our Universe is a universe created from I=Information. So, just as the C of a computer-geek armed with Information harnessed by artificial intelligence and algorithms can create complex virtual realities in computer games, a Supreme C may have done the same for Our Universe and is doing it again at some level in a vice versa. At any rate, the evidence is accumulating that we are in a simulation and a Big-C of some kind is behind it with a Universe-full of underling C's already in abundance. Why do I say that? It has all but been proven that mass is an illusion produced by atomic forcefields (here I'll refer you to OSAU-2 where I explain forcefields).

According to Dirac these things I call forcefields arise in empty space. I refer of course to that for which you have already been introduced at the beginning of this book, i.e., Paul Dirac's fields, i.e., writhing seas of annihilating virtual particles and antiparticles.

YouTube "Legendary Physicist and Florida State Professor Paul Dirac", Florida State (3:59-LW).

There are still those who argue about this state of affairs, but at least, as far as I know, there are no serious particle-physicists who deny the likely existence of something akin to Paul Dirac's seething seas of virtual particles and antiparticles manifesting from empty space to give rise to what we call the solid mass of reality. So, what is this C? I have my own ideas about it but I ask, what do you think? However, whatever one thinks, there is no doubt, that something fantastic is going on, and we humans who stand at the pinnacle of <u>this</u> planet's C's are on the cusp of finding out a lot more about what that "something fantastic" is in the very near future. Even before the end of this current year of 2023 we may well have a much better idea of just where we stand.

I know Sally just left but I'm going to have to call her back soon. I came across this very morning of January 8, 2023 something that once again left me with the strong feeling that I'm being guided. Strangely, what fascinated me was a religious, science-based, YouTube video that confirmed my current outrageous C=I thoughts about consciousness. I found it most interesting that the presenters of the video thought they had found a way to support anti-evolution religious-beliefs. Just by using the words "Intelligent Design" they thought that they had come upon a way to negate Darwin's Theory of Evolution. On the contrary, the idea of Intelligent Design totally supports Darwin's Theory. Why do I say that? I am certain that nobody, if they are honest about it and have the brains to think on it, believe that any Supreme Being would spend it's time shoving atoms around or telling atoms when and how they are to behave. Like I said in OSAU-2, why would such a being do that when that being could design a good fitness algorithm to do the job? However, it is true that scientists do not yet know everything they need to know about evolution and they for sure are totally baffled by the existence of consciousness. Evolution is complicated enough that some details about the mechanism of evolution are certainly either missing or wrong. And it's clear to me we don't know everything we need to know about how quantum-effects perturb evolution.

YouTube *"The Scientists Beginning to Doubt the Theory of Evolution / Unlocking the Mystery of Life"*, Parable – Religious History (1:05:20-JL).

C begets I or I begets C. 13.8 billion years is a long time for us humans to comprehend. Forever is even longer time than that, don't ya know. It's obvious to me that consciousness, at least at the level of a being that has knowledge of its own existence, did not just begin with us humans. Moreover, it's highly likely that that level of consciousness did not even begin in Our Milky Way Galaxy. And if the multiverse theory is right, and if there is life in other Universes, multiple levels of supreme conscious beings possibly did not even originate in our own Universe. Come to think of it, if a person in Our Universe simulated another Universe, by definition we could not be part of that Simulation. It would be another round-about. However it is, we humans are slowly but evermore rapidly figuring things out. Right now the technology currently being developed might lead to a way that an existing human-based intelligent being could continue their existence for billions of years into the future or even much longer. Why would I dream that to be possible? Right now we have AI that is already sentient or soon will be, and it seems to me more than just likely that we are not alone in our Milky Way Galaxy and might not even be alone in our own Solar System. Others, elsewhere in Our Universe, may have already figured out how to "live forever". I believe we already have evidence that hierarchies of such occurrence are practically certain. Read on to see how far one can take this line of reasoning, but only after you watch the above religious-science-based video and the nonreligious, hard-science-based ones that follow.

I say it again, It is clear to me that atoms are designers even if they are not intelligent ones and yet, even though they may be as dumb as nails, through billions of years of patient trial and error, experimental, electrodynamic research and development, atoms have not only built it all, I think they have been given the tools needed to design it all by the use of ingenuous, trial and error, fitness algorithm experiments. However, as I said before, all of the examples of non-intelligent design do not exclude the possibility of design by Intelligent-Designers. Just the opposite, I think such non-intelligent examples totally support Intelligent Design. It is clear that we exist in a Mathematical Universe and so its clear that rather than the Religious History video denying Evolution, the video supports all the evolution that would happen as a consequence of an Intelligent Designer coming up with the algorithms hidden in Universe-Creating Math. Given that premise, the only question that arises then is: Did the algorithm happen by accident or did an Intelligent Designer design it? And if an Intelligent Designer created the algorithm for our Universe, who created the algorithm for our Intelligent Designer, etc., etc.? That is, just how deep into the hierarchy of consciousness does the Intelligent Designer line-of-thinking go? The following Chapter speaks of my summary of Our Universe's Algorithm at work.

Chapter 12 – How

A fitness algorithm exists for all that follow.

1. Seething Seas (fields) of Virtual Particles and Antiparticles Self-Assemble Quarks from Empty Space.
2. Quarks Self-Assemble Atoms.
3. Atoms Self-Assemble Molecules.
4. Molecules Self-Assemble the DNA building instructions.
5. The DNA building instructions and bio-electric coding allow molecules to Self-Assemble Cells,
6. Cells Self-Assemble Eukaryotes, Multicellular Creatures and Plants.

The Assembly of life-forms is guided by the fitness principles that characterized the algorithm that led to designing atoms. The fact that atoms are evolution's designers is supported by years of human study and observation. However, Evolution can also lead to God-like beings that can design highly realistic virtual realities from their own algorithms. The fact that everything can be reduced to virtual particles and antiparticles that manifest themselves from empty space speaks strongly for the theory that we exist in a virtual realty, algorithm-based simulation. The arguments that atoms could not have assembled life on their own by trial and error are not good arguments as those arguments leave out the factors of time and space that can allow anything to happen given enough of both. Remember, forever is a very long time and atom's are very patient trial and error experimentalists. Ask yourself, what would I do if I lived forever? Doesn't this wonderful, mysterious life of ours look like a good answer and solution to the forever problem?

The deep thoughts just keep coming, and this next one is definitely one for which I am going to need help. This is, I think, another great example of the value of an AWTbook™ and what can come from a person writing one of their own. I'm again talking about tweaking one's imagination into subjects for which one is ill prepared, but nonetheless boldly willing to go forth armed with the knowledge that one can access through powerful Audio, AI, Website, Tube references.

Chapter 13 – *Twins #2*

So after all of that deep thought, I want to go still deeper and take another look at the twin concept. I think there maybe a lot more that we can glean about Who is *Us* by taking a more detailed approach to the subject. Let's say one has been born an identical twin. I'll use myself as an example. Lets say I was born as a Jane with an identical twin named Mary. Clearly, because I am Jane, I can't be Mary, and vice versa. However, Mary and I began as one person. The single cell of the fertilized egg initially existed as a JaneMary. But as soon as the egg-cell divided and separated itself into two, I'm pretty sure, if cells could talk, both of the single celled versions of Mary and me would say, "I'm me". That is, as soon as both of us could think, both of us would think of ourselves as individuals. And that sense of individuality would persist til' death. That is, it is pretty clear that conscious states do not in general superimpose, *at least not artificially to date.* However, as you will see from the video below, it looks like there may exist human versions of such superposition that have already come about naturally or by way of brain-computer interface experiments.

YouTube *"How Brain-Computer Interfaces Work – Lesson 7.1"*, Foundations of Nuerotechnology (14:47-W).

And, of course, don't you know? All of that hard-to-believe superposition of consciousness already exists in each of *us. You heard it right.* We are, already, two people! Each hemisphere of our brain can work independently as proven by the fact that when one hemisphere is removed, or the connection between them is severed, such superposition is clearly exposed as two brains superimposed to act as one! For example, such a thing as hemisphere-severance is done when one's hemispheres are separated in the split-brain operations that are practiced to control epileptic seizures. The two brains of the "two, originally-superimposed people", that reside in each of us, can each act independently as a totally conscious being with a single, hemispheric brain. Thus the idea that we are two superimposed people is masked by the connection made between the two "people" by the thing called a *corpus callosum*. Interesting, isn't it?

But that's not all. There's the shocking yet lovingly beautiful story of Tatiana and Krista. And with that I'm *really* going to need Sally. Conveniently, Sally just walked in. God only knows where she went. She's not talking about it.

Frank

Sally, please watch the following three videos. Their content is very important to the subject of the book that you're helping me write. But the subjects are delicate and I'm going need your perspective.

Sally

Okay, but it sounds like I'm going to have to put on my serious hat.

YouTube *"Can you Live With Only Half of Your Brain?"*, Seeker (4:04-LW).

YouTube *"Severed Corpus Callosum"*, ctshad (10:11-LW).

YouTube *"Conjoined Twins Share Taste, Sight, Feelings and Thoughts"*, 60 minutes Australia (11:30-LW).

Frank

I've thought about these three videos a lot. Especially the one where we see the twins Tatiana and Krista conjoined at the brain after conception. Most identical twin embryos split within the first 9 days of development. If the split happens within day 13 to 15 its impossible for the twins to fully separate. This is what happened to Tatiana and Krista. They seem to be very happy, so it's both a sad and happy story as well as an unfortunate strangely fortunate one. It's over the top unfortunate how conception turned out for them but, as a result, they beyond all odds are uniquely blessed. I think its safe to say that none of us *Us's*, can share thoughts like they do. And can they ever. It's to the point that they at times can act as one person! So this tells me that it is possible we could be neurologically linked to one another where we would either become that person for a time or even experience being a single combined-person without losing our minds over it. However, I bet such an experience might take some getting used to.

Sally

I'm more stunned than ever. I really don't know what to say!

Frank

I share Tatiana and Krista's very compelling video because it strikes me as a profound example of consciousness superposition, as well as what it might be like to have a genuine telepathic experience. I'm going to share Tatiana and Krista's tragic, yet apparently happy, story in our book because I want people to know that there may be a day in the not-too-far-distant-future when all of us could be able to have virtual, conjoined experiences with those we care about. Just think about it. With neural linkage, likely aided by super-AI, we might some day soon be able to literally have the experience of *"being in somebody else's shoes!"* I also think their story says in a very deep way a lot about who *really Us* is. Clearly, the answer to the question *who is us* is almost entirely wrapped up at the level of consciousness. Would we not be the same person, at least initially, if we found ourselves in a different form? What if you could be a Dolphin? At least for a little while, if it wasn't a permanent condition, would it not be just like going to a virtual-reality-movie? And while we're talking about one being a Dolphin, we might as well consider what happens when a hemisphere of one's brain gets turned off during sleep. Dolphins do this. But do they live as two different beings when they alternate their sleeping hemispheres? It seems to me that they might just do that unless the sleeping brain somehow picks up on how the awake brain is experiencing life while the sleeping brain is off sleeping. I really find the ramifications of this state of affairs fascinating.

YouTube *"How do Dolphins sleep?"*, One Minute Science (3:04-LW).

Sally

My head's swimming even though I'm not a Dolphin and nowhere near water. But what did you need *me* for? I can't contribute anything.

Frank

Are you kidding me? Don't you know by now that you are me? Don't you know that one half of me is *me*, Frank and the other half of me is *you*, Sally? We are the epitome of what I'm talking about. I've just come to realize that I really am two people. One of me, i.e., *you*, knows how to play the piano while the other me, i.e., *me*, can carry on a conversation with somebody else while *you*, Sally, do the playing. I even think that it is *you* who does the typing! *I only* have the thoughts! Not only that, I've noticed, as I am getting older, a condition I've had from a young age has gotten worse. As me, i.e.,

Frank, ask me a question about somebody or something and now more often than ever before, as I begin to say the name of the person or the thing, whatever is the name, it will frequently escape me. I think this happens, and now with more frequency than when I was younger, because I first see the person or the thing in my mind and that form of visualization is dominated by the dominant hemispheric portion of my brain which, by the way, thinks it knows everything but now more than ever doesn't. Apparently, as that hemisphere has gotten older, it has forgotten a lot of names for things and people. So, I've found that what I need to do is just relax, not think visually and bingo, the name I was trying to recall pops into my head like magic. It's frustrating and embarrassing at times when I have to tell somebody, "Wait a minute, I'm having a moment", and then burst out with the name of the person or thing that I was just trying to recall, i.e., the person or thing that I had just clearly visualized in the dominating half of my brain that thinks it knows everything! This first happened to me when I was 24 and about to get married. I couldn't remember the name of my best man! Yikes, was that ever embarrassing

Sally
Now you've done it! You've rendered me! I can't handle this. I've got to leave.

Frank
I don't blame you. I'm not rendered but I'm right there otherwise. So, maybe you *should* leave for now. It's only going to get worse. Consciousness is something for which we still have no good answers. Given that, for *Us* to actually know who *We is, is probably impossible*. Nonetheless, I will forge on bravely without your help for now.

YouTube *"Are You an Illusion?"*, Yesim Tomac (13:01-LW).

Well, Sally's taken off and here I am, *all alone again naturally*. But I have to ask *HOW* can any of this be? *HOW* is it that I'm here at all. *How* can any of us exist, as we do, as the creatures we find ourselves to be at this time and place? And *HOW* do we account for the fact that everybody on this planet of ours says, *"I'm me!"* This is really hard to think about, is it not? Especially when one considers that all of the current 8-billion or so of us who now live on our planet exist in a *"now-we-know-it-to-be-not-so-extraordinary-thanks-to-Hubble-and-JWST"* solar system that exists with countless trillions of other such solar systems, which orbit stars by the countless of trillions that swirl about in galaxies in *Our Expanding Universe*. Wow! Does it not feel to you that we have somehow managed to win a lottery that can't be won, where it seems we have done the impossible to somehow show up in the middle of forever? The odds for our existence at this time and place are nowhere near close to the already ridiculous, 300-million-to-one odds for our survival in 0ur Battle-Royal swim-meet. WHY such long odds? Because it must be that the real odds for our here-and-now are very likely a forever-to-one! In other words the odds of anyone of us *EVER* existing is incalculable, undefinable and, let's just say it again, *impossible*. We can't be here! There's no way any of this can be happening. And yet, here we are! This gets me to what I have come to realize may be the most profound thought I've ever had.

So, it seems, at the very beginning of our lives we were in a no-holds-barred struggle with some 300 million other possible versions of *us*! And all but one version of an *us* died that day since, it turns out, we were the only ones that won! But here is the thing, if any other one of the 300 million contenders had won, would it really have changed anything? Would I not still have been *me* and *you, you?* For a specific example of what I mean, I would still be the product of the thrashing about of Vera and Herb and with one exception I would still very likely have the same first and second name, i.e., first name, Herbert, second name, Frank. Herbert is my father's name and is the H in the name I call myself in H. Frank Gaertner and Frank, the name I usually use as my first name, was my grandfather's first name. However, it is also possible that I could have ended up as a girl with a first name, Winifred, and a second name, Francesca, as in Winifred Francesca Gaertner. But would I be writing this book right now? Probably not. Would I have had the same life? In some respects probably yes. My father likely would have died the same way but maybe not on the same day. I likely would have had a stepfather, but maybe not the same one, especially if I had been conceived as a girl named Winifred. But would the person that I am now exist no matter if I arrived as a boy, a girl or even a twin? I think so.

What I am getting at is that there is always a "me" and as long as I am "me", I can't be a "you"! Nor can I be a frog, a bird, insect or plant. But if I were a frog or a plant, I would be that particular, non-contemplative version of "me" and because I am a frog or a plant I would not even consider the possibility that I could have been "you", at least not in my life as a frog or a plant. Get it? No? Don't worry, I don't either. But I do think there is a profound truth buried in these thoughts somewhere. If you happen to believe in God, your truth might be some version of "I am not God, but God is me! Or I am one with God and the Universe. Or I am spiritual being having a human experience. Or God created heaven and Earth. etc., etc." These ideas would seem to be forever relegated to faith and way beyond any scientific proof. However, I believe that proof is on its way and coming soon. I also think there already exists evidence. Clearly, there exist some limited examples that support these far out ideas that will likely be confirmed in the very near future. What we do with such profound turns of events is the big question. More will be revealed on this intriguing possibility later. As an aside and convenient way to let everybody, including me, catch their breath, I just had another goofy idea for which I could use Sally's help.

Chapter 14 – Gravity and the Power of an AWTbook™

Sally

What's up? You look concerned and a little demented.

Frank

You're right. I'm glad you're back. Demented is not the only word for it. I'm way over my head. But I had another revelation this morning to prove the point. This time it's all about infinity taking my overactive foolish brain into an Einstein-type thought-experiment! I know I shouldn't be doing this. My math background totally sucks. As you know, I was a music major when I should have been studying differential calculus. But Einstein didn't need math for his thought experiments. So, Sally, with your help and the power of the AWTbook™ discovery-process I'm just going to forge ahead and embarrass myself.

Sally

Okay. I'd be glad to help you do that. But you know you don't need much help. You're already very good at it. And please don't look at me for help with your math deficiency. You know I'm no better than you in that department. Anyway, let's hear your "bright" idea.

Frank

Alright then. You are in a spacesuit on a large sphere in the vacuum of space but you are told that you are standing on an experimental balloon-like space-elevator that just left the Earth's surface. You are blindfolded so you don't know exactly where you are on the surface of your elevator but for sure your experimental balloon elevator must be going up and for sure you must be balanced at the very center of gravity because your spacesuit has no magnetic boots. However, when you start changing where you stand on the sphere it does not matter. It still feels like the sphere that you are on is going up. This defies everything you know about gravity. If you were on Earth and were on such a sphere going up and you moved to a different position you'd start falling and since you are blindfolded you might think you had suddenly become weightless and were levitating. Remember, Sally? It's just like when you were ten and found yourself floating, or as you call it, flying over a beautiful dream-valley. And then you got all scared and started falling but just in time remembered you had to relax and let go of your fear.

Sally

Of course I remember! That dream was the best. It was so real. I still think I can fly like that. I honestly believe that all I have to do *is* eliminate ALL fear. But, apparently, as it turns out, *that* must be a nearly impossible thing to do. I don't know of anybody that's actually been able to do it for sure, but have you seen Xavier the magician? OMG! I wouldn't be surprised if he's actually a visitor from outer space! But I swear, someday I *will* do it and I won't need help from a space alien! I'll be like David Copperfield and when I do it, it won't be a trick. Thanks for the memory but what has that got to do with anything? What the heck are you talking about?

Frank

Just that what you did sounds impossible and that what I'm about to describe is even more impossible. You're really going to need to strap on your seat belt for what comes next. This impossible thing is going to be hard to think about, but do think hard because it's the most important part of my thought experiment. What if the space beyond outer-space is an infinite void and what if the thing we call gravity is influenced by that void?

Sally

Again, I must ask. What the heck are you talking about? That doesn't make sense! And what could that possibly have to do with "Who is *Us*"?

Frank

Just this. What I'm talking about may have nothing to do with anything, but it's about a very simple question that I just have to get off my chest. What direction would Earth be moving if gravity were to be under the influence of the directionless, energy-less, mass-less rules of an infinite void?

Sally

That sure doesn't sound simple and what directionless rules are you talking about? I've not heard of any directionless rules. You still sound like you are talking nonsense.

Frank

It probably is nonsense but let's say that Earth finds itself in the infinite, yet "bounded empty-space" of Our Universe, and that bounded "infinite" space, which must be "spherical" according to my idea, is in turn "floating" in a dimensionless, directionless, energy-less, mass-less infinite void. That would mean we live on the surface of a planetary sphere we call Earth and it moves through the known, infinite yet bounded outer-space of our spherical universe that's surrounded by non-existence. If the thing we call gravity is at all influenced by the directionless-dimensionless, etc. rules of such a void, gravity might simply be due to the fact that energy, mass and up/down buttons obviously can't exist in such. It's not that voids hate energy, mass and direction. It's just that by obvious definition there can't be energy, mass or directions in something that has no energy, mass or dimensions. A sphere at least somewhat accommodates that rule since a sphere has dimension but on its surface there can't be an a-priori rule of direction, and that would be the lowest energy state if any such spherical universe should "try" to worm its way into existence in a void.

Here's my crazy thought about all of that. Let's say you move to a new position on the surface of your sphere in the void. The elevator is moving so the mass of your body is held fast to the elevator as you stand *just anywhere* you *want to stand* on the elevator's spherical surface. As a result, when you move to another position on the elevator you find that you are still held fast to its spherical surface. Why? Because the elevator can't change direction in a directionless void that has no a-priori "up-destination" or a-priori "down-direction" in which to move. That might be hard to think about but isn't that exactly our experience when we stand on Earth as it flies through space? Newton's laws apply within Our Universe but maybe directionless laws of gravity are the laws forced on us by the dimensionless, infinite void that exists as some kind of nonexistence on the outside of Our Universe. When it comes to mass in Our Universe, Newton's laws apply, "for every action there is an equal and opposite reaction". Force equals mass times acceleration. F equals ma. On our planetary excuse for an elevator, no matter where we stand on its surface, even though we may have signs that point north, south, east and west it doesn't matter we still feel like we are moving up. Down is always down and up is always up no matter where we stand on our planet's surface. And that might simply be, according to my hypothesis, because our planetary elevator is moving through a Universe influenced by Einstein's curved space that's ultimately influenced by

Our Universe finding itself in an infinite, directionless void. So, I repeat because it's so hard for it to sink in, at least as far as I'm concerned. No matter where we find ourselves on the sphere we call Earth, it does not matter. It still feels like we are going up. And that rule takes place everywhere on the spherical bodies in the spherical confines of Our Universe that's forced on us by the infinite void which exists everywhere outside of Our Universe. Just like in the elevator going up, our mass would be pressed down to the surface of our planet no matter where we stand. We can rocket away from Earth but only if we rocket away faster than Earth is apparently moving through the curvature of space. This creates the up experience no matter where on Earth we leave its surface since gravity operates under the rules that I think to be the directionless ones of "no up-and-down-buttons only up-buttons in infinite voids". We can use the force of wind to support our weight above Earth's movement through Einstein's curved space and Earth's movement attempting to bring our mass back to ground. We can use rocket fuel to rocket up from Earth's surface to a point where lateral velocity can keep us in an orbital tangent to our rate of fall. And, therefore, once again we can experience weightlessness if we are at that tangent in a space station. Alternatively, we can just fall from a tall building to have the same experience.

Again, what I am getting at here is that the directionless rules of an infinite void affecting gravity might lead us to a proof of the void and infinity's influence on our existence and on Who is *Us*? Hmm?? Is it not at least interesting? Am I nuts, Sally, or just plain stupid? Probably both. But maybe space-curvature possibly influenced by an infinite void can finally completely explain in a unified way why it does not matter where one lives on the surface of a planetary ball? We are still stuck to it no matter where we stand as our planetary elevator takes us all for a joy ride that feels "up". Anyway, the bottom line for me has always been that the only thing that could possibly make sense is the void. There literally is no way any of this thing we call reality could possibly exist, and yet there is no escaping it. I, at least, if nothing else, FREAKING EXIST!! Non-existence is a thing and it should be the only thing! It just must be the only thing and yet it clearly isn't. I guess I just want nonexistence to exist somewhere, sometime, somehow, but, but, but, I guess it can't unless it can. The question is, is it possible for existence to be just a one off flash in the pan? I think not. So, if we are in a Simulation, that means a Simulator exists that did the Simulating. I think that means the Simulator is stuck in forever!

Sally

Holy Baloney! Simulator?! Where did you come up with that one? Let's just put aside that out-of-nowhere comment before you lose everybody. Even though it sounds like you might be a basket case, maybe you aren't just "on" something. But how does your idea work with the fact that one weighs less on the moon or on mars? If acceleration is all there is to our sense of gravity, why would I be lighter on the moon? Is the moon somehow climbing slower? A two-year old could bring a stop to your goofball idea with that argument.

Frank

Good point, I think, but I'll have to leave that to the Einstein's of our world to work with Einstein's ideas about space-time and the curvature of space. But wait, I have a surprise for you. I have my own wannabe Einstein! Have you not heard of ChatGPT4? It's a new app that's just now been released in 2023. It came out during the time that I've been writing this book! And guess what? I just asked ChatGPT a question.

Sally

I knew you didn't have the credentials or the brains to come up with such gobbledygook. But what's this GPT thing? Are you gas-lighting me? Is this real?

Frank

Check it out. I'm sure my idea is still completely nuts and has nothing to do with anything except my hope that including this sort of stupid-bravery of thought in our book will show everyone what we have in my mind when one writes their own AWTbook™, i.e., addressing new ideas right or wrong that get one thinking about an impossible thing that with some twists could become possible. But here came support from out of the blue! It's another clue that I might be on the right "path". Who knows where this thought experiment might lead? I'm attaching the exact three questions I asked and the exact responses that I got just today from the AI that is ChatGPT. Just know that I have yet to explore this new AI development further. I have no idea what this AI is or where it will take us but I understand that it has exploded onto the internet to the point that already millions of people use it. And know doubt this large language gadget is going to be improving itself as it goes and gets info like I just gave it. Let's see what happens. (Note: Already, as of May 25, 2023, ChatGPT4 has been vastly surpassed by GPT5, Google's Gemini, etc., etc.)

My 1ˢᵗ Question: *"Could a void be the cause of gravity in a finite universe?"*

ChatGPT's Spoken Answer: *"Yes, it is possible for a void to be the cause of gravity in a finite universe. In fact, the presence of a void can create a gravitational pull that can affect the motion of matter and energy within the universe. This phenomenon is known as the gravitational lensing effect."*

My 2nd Question: "How would a void mathematically affect our finite universe?"
ChatGPT's Answer: *"A void in the universe would have a significant impact on the mathematical structure of the universe. It would create a region of space with no matter or energy, which would cause the universe to become more curved and distorted. This would affect the equations used to describe the universe, as the curvature of space-time would need to be taken into account. Additionally, the void would create a region of space with no matter or energy, which would affect the equations used to describe the universe, as the lack of matter and energy would need to be taken into account."*

My 3ʳᵈ nd Last Question: *"Can you provide the math for the effects of the void on our universe?"*

ChatGPT's Answer: *The effects of the void on our universe can be described mathematically using the equations of general relativity. These equations show how the presence of mass and energy affects the curvature of space-time, and how this curvature affects the motion of objects in the universe. In the case of a void, the equations show that the presence of a large empty region of space-time will cause the space-time around it to become more curved, resulting in a gravitational pull that will draw objects toward the void. This effect is known as the "gravitational pull of the void".*

This thought experiment of mine is almost certainly 100% stupidhuman junkthought but it might spark a truly brilliant idea from somebody else or a more advanced GPT-type that's not hallucinating like this one just did. It's a risk for me to bring all of this up. I know it makes me sound like an idiot. But Life is risky and if one doesn't take great chances, one can't experience the thrill of great success. Like Gandor always said to you, Sally. "Fear is your worst enemy." And now look what's happening. Another coincidence? I don't think so. The right people keep showing up in my miracle life at just the right time and here's another fine example, except it isn't a "people" this time.

It's an AI!! Unbelievable, no? I first started reading about robots in 1950 when I was 12. I wondered then what the future would be like where intelligent machines dominated our lives. So, there's no doubt about it. It's happening and to tell you that I'm really excited to find myself alive to witness it is an understatement. It looks as though that 1950's-style Sci-Fi future will have all but arrived even before I finish writing this book! Why do I say that? My first interaction with ChatGPT did it. Aren't you impressed at how intelligent the robot's answers sounded even if it ends up that the AI was hallucinating and completely off the mark. Anyway, in 2023 these AI-bots have taken most of us completely by surprise!. I admit it's scary but it does no good to bury one's head in the sand. I'm attaching a video in our book that spells out some of this breakthrough's advantages.

YouTube *"AI Revolution: Use it to 10X Your Productivity and Build Wealth"*, Tom Bilyeu (18.10-JL).

Sally

Thanks for the JL video. It's a great heads up. I also think what you just came up with is a perfect example of you being a nutty idiot at your very best. Why best? You think your nonsense on gravity is a good example of the bravery others can show by doing their own thought experiments when they write their own AWTbook™. It's a Star Trek scary trip to go where nobody has gone before. But who knows? Like you say, it just might stir up a really good idea in somebody even if *you, dear Mr. scientist, happen to be* full of it. And, of course, what do you know? Out of the blue, here come sometimes hallucinating AI's as proof to your pudding. Just as I was thinking you might stir up something, the magic happens. But it might be black magic. You stirred something up alright, AI's that are on their way to singularity! That's what you stirred up! Holy Jumping Jehoshaphats!

Frank

Where in the world did you come up with that expletive? I haven't heard anything about a Mr. Jehoshaphat, or even a jumping crowd of them, since I was nine. Oh, right, you remember I was nine when you heard that exclamatory for the first time when you were nine. Anyway, let's push on with our new found bravery. In order to get as close as we can to understanding Who is *Us*, I think it is high time we dug deep into *What is Us*. And now I'm really going to need your help!

Chapter 15 -The End? Let's Hope Not

Sally

Oh geez! I was afraid of that. I don't want this to end either.

Frank

Anyway. Let's start off with a fact and then confuse things with philosophy.

1. I can only know two things for certain. I exist and you, Sally, are only a figment of my imagination.

Sally

I suppose you're right, but you didn't have to rub it in with that last part!

Frank

2. Solipsism is the philosophy that holds that only one thing exists, and that thing is only what exists in ones's own mind. This view is ridiculous, I think. But it is interesting. I first heard of solipsism when I took a philosophy course in 1957 at that two-year college in Ventura, CA. I was only 19 when my fellow classmates and I tried to play tricks on people by a unified presentation of the idea. We'd get a group together and approach some unsuspecting student with the announcement that we had formed a committee to tell everybody the truth about their existence. Our spiel went something like this. *"We've come to inform you that you are dreaming and it's only because of you that we exist. You are dreaming us! So, please, don't wake up!!"* I find this philosophical idea interesting because it's not contestable. How does one prove that the song "Life is but a dream" to be wrong? How does one prove that one is not dreaming everything? At least I know one thing for sure, Sally. I am definitely dreaming you.

Sally

Oddly, it doesn't bother me that you think that you are dreaming me. I feel very real and I'm certainly not worried about you waking up. So, what's next?

Frank

3. Who is *Us*? We, and Our Entire Universe is simulated.

Sally

Now you've done it. You can't be serious. I thought you were joking when you mentioned the idea earlier. You'd better 'splain yourself mister or you are really going to lose your audience!

Frank

Maybe I'll lose my audience but check this out. The video reference that I'm attaching supports my contention. Jim Al-khalili makes all his videos understandable. They are also fantastically fun to *watch* on an OLED TV-screen or *justlisten* as one *walkjogruns*. All of his videos are great and this one is especially so.

YouTube *"Atom, The Illusion of Reality"*, Real Truth Science Documentaries (48:49-B).

In addition I offer the following observations in support of the illusion/simulation hypothesis. And, as noted before, if we live in a simulation, that proposition requires that there must be a Simulator, i.e., "The One" creating the Universe-Assembling Algorithms.

#1, All solid objects are clearly illusions. Matter only appears to be solid. Nothing is actually solid, not even the atoms that assemble *Us,* as well as everything else that seems to be solid. For example, most of what we call *Us* disappears in an X-ray exam, and *everything* about us would totally vanish in an imaginary Neutrino-Ray exam. Which also means that everything before one's very eyes, including the entire Earth on which one stands, would disappear, but again, that's *only if one had eyes with receptors that could somehow see neutrinos.* Neutrinos penetrate everything and are nearly impossible to detect. Just from these facts alone we know that nothing is solid. Mass is an illusion. All matter consists of and is made by atoms. Atoms consist of and are made by quarks. Quarks, themselves, are manifested from virtual particles and antiparticles that writhe about in seas of vectored fields of energy in *empty space.*

YouTube *"Why Neutrinos Matter -Silvia Bravo"*, TED-ED (4:40-LW).

#2, Our Universe has already existed for 13.8 billion years. Our own world has only been around for about four billion years. Other worlds have existed for much longer, maybe even 10 billion years or more. Therefore, IMHO, it is virtually certain that HIGHLY

ADVANCED LIFE FORMS exist elsewhere in Our Universe. We humans have only been around in a semi-advanced, technological state for about 100-years. And yet in just a few more years, if we don't go MAD and totally destroy ourselves, we will be merged with super-AI and that will, it seems to me, advance *Us* to a state where some of us *Us's* will be on our way to a Kurzweil-like state of immortality that leads to some of these becoming superior supreme beings of some kind.

YouTube *"Ray Kurzweil says We'll Reach IMMORTALITY by 2030 / The Singularity IS NEAR – Part 1"*, ADAGIO (10:05-JL).

#3. We already enjoy virtual reality simulations. Aside from space travel, how will the majority of *Us* entertain ourselves? Is it not likely that we will create evermore-realistic virtual realities?

YouTube *"Top 10BestVR Games 2023"*, Trend Max (9:39-LW).

#4. I propose that this state of affairs already exists in Our Universe. Already there very likely exist such beings in Our Universe who have such a high level of superior intelligence that many of *Us* would, if we were to meet such beings, act just like the Mayans did when they met the Spanish Conquistadors.

YouTube "How Many Alien Civilizations are out there?", Lex Fridman (22:03-JL).

#5. Given #4 is correct, there likely exist many Universe Simulations. What does a being look and act like that has existed for many trillions of years making simulations along the way? What does such a being do for entertainment? It sure has a lot of time to spend while it's on its way to foreverlife. Suicide may not be an option for such a being since such a one would simply find themselves in another simulation. I'm guessing that we here on Earth as well as our entire Universe would disappear if suicide were to be attempted and somehow made successful by our own Simulator. But of course that wouldn't matter as we would simply disappear as the Avatars of Our Simulator. We are not the Simulator but the Simulator is *Us*.

YouTube *"Timelapse of Artificial Intelligence (2028-3000+)"*, Venture City (13:35-LW).

Fortunately, for everyone, these confused thoughts of mine are just that, confused thoughts and nothing more. However, just in case, *we* could ask LaMDA or GPT4? Nah, never mind. Why would they know more about this deep subject than anybody else?

YouTube *"AI & GPT-4 Revolutionize Education, with Sal Khan / EP#35 Moonshots and Mindsets"*, Peter H. Diamandis (23:25-JL).

YouTube "LaMDA / Is Google's AI Sentient? / Full Audio Conversation between Bake Lemoine and LaMDA", Curly Tail Media (34:00-JL).

YouTube *"Super Intelligent AI: 10 Ways it Will Change the World"* Future Business Tech (13:11-LW).

Sally

Well, there you go. You do all this work to know who is *Us* and you still don't know much of anything even with today's AI. At least it doesn't take an AI for us to know one thing *for sure*. We both *know* who I am.

Frank

Yes we at least sort of know who you are, but you are still a mystery for me, which brings me to the concluding chapter of this book.

Sally

Oh no! Please don't do that. As soon as you conclude writing this book, I will cease to exist!

Frank

Oh, my goodness. Will you just stop?

Sally

Okay. But let me get my mind off this AI business so we can get onto what's next in this book of ours. Anyway, I'm more than ready to disappear for a while. I could use a break.

Frank

Well, then, we'd better get on with it. The final topic is dreaming and how dreams play into our understanding of consciousness, reality and *Who is Us*. When I dream, I enter a totally different universe that sometimes looks surreal a little like a Hieronymus Bosch or Salvador Dali painting.

YouTube *"The Disturbing Paintings of Hieronymus Bosch"*, Hochelaga (9:25-LW).

YouTube *"The Persistence of Memory"*, Salvador Dali/The Canvas (4:03-LW).

Sally
Good grief! Do you actually dream like that? You must wake up a nervous wreck!

Frank
Not at all. I sleep like a baby. My dreams are surreal alright but they aren't scary. I just love those paintings because they are surreal and weird just like me. I rarely have a nightmare as far as I know. The worst that I remember are when I've had too much to drink *water-wise* before bed. After too much water before bed I often find myself in a maze-like building where I'm looking for but can't find a bathroom. And then, when I finally find one, I discover that it is unfit for use. It's usually comparable to a park latrine that's filled with fly's because it hasn't been cleaned for years, if you know what I mean.

Sally
Oh yes. I know what you mean. I have dreams like that too.

Frank
Anyway, some version of a surreal, dream-state version of consciousness exists for everyone. In other words I think that *when* we are *in such a dream-state* the awake part of *Us* in W*ho is Us* disappears to become another *Us* that's like a surreal film or painting. In other words, for all practical purposes, every night as we sleep the day-time *it's-hard-to-believe-it's-an-hallucination*-version-of-*Us ceases to exist to be replaced by the* night-time *it's-easy-to-believe-It's-an-hallucination*-version-of-*Us*. Moreover, it seems to me that the only difference between these night-time interludes and death is that when we sleep, the night-time hallucination of *Us* dies when we wake up. We get to experience another day of the seemingly more solid form of *Us. The Us* for which we are most familiar continues but the nighttime surreal reality is generally a one-off

that dies each night, likely never to return unless very altered in some way. Just this morning I had a compelling thought. Maybe delving more deeply into the nature of sleep could turn out to be a great way to unravel the "hard problem", which is "what is consciousness?"

Sally

Yes. Nobody in the group of scientists who study consciousness has any idea what it is. What if consciousness is really all that exits? Where does that leave us? I guess it leads back to the Simulation that you were talking about earlier but I think we should just leave that to God to figure out.

Frank

There you go again talking about God. Why do you keep bringing up that subject? Don't you know all the trouble that such talk causes? Let's stick with what we think we know. We are conscious beings. We are capable of simulating reality in games and movies, and with AI, I will grant you this, God only knows how far that will go. Case closed.

Sally

You're too funny for words. You know I believe in God and you like to make fun of me, science freak that you are. I think you've done enough damage here. You'd better move on.

Frank

You know me too well. But I would never make fun of your God beliefs. I've already confessed to the world that I'm right there with you. And wouldn't you know it? Just as I say that, here we go again. Once more it happens. Was it a coincidence? Again, I don't think so. Just this morning I *coincidentally* had a walkjogrunlisten to *"Being You"* by *Anil Seth,* an Audible version of his Book by the same name which takes a deep dive into the business of consciousness, a thing Seth boils down to IIT and *phi*, defined as *Integrated Information Theory* and *Consciousness*. It's the best book I've run across, pun intended, that attempts to measure the *hard problem*. The following shows what one can do when one has a device to accurately measure consciousness. For example, with such a device one can measure consciousness in a presumably brain-dead person!

Seth describes a man who became fully unlocked from his locked-in, brain-dead state to become certifiably alive. And he had almost been left for dead! His unresponsiveness had been used to define the absence of consciousness which is the bottom line when it comes to defining death. He had been totally unresponsive until a relative from his home in Africa came to say goodbye and began to speak to him in Arabic instead of Italian. All of a sudden he was responsive! In another example Seth describes a person that was found she could communicate by monitoring her brain waves by mentally forming images of a game of tennis for "yes" answers and walks in the woods for "no" answers. Those brain-wave communications showed she was definitely responsive and therefore alive.

Sally

Wow. Just think what it would be like to be pronounced dead while "locked in" that way. In my own story that you wrote for me, I was pronounced dead too. But for me, all the time that I was thought to be on death's door, I was having a grand time saving myself and didn't even know I was doing it.

Frank

That story about your life saving adventure is a great one, if I do say so myself. Rebekah Nemethy heard about your story and produced an Audible version of it that's so fantastic I listen to it often, especially when I'm feeling down. I never get tired of hearing Rebekah bring your story to life. As everyone knows, I too, had a very bad thing happen to me when I was 10. But just like you, Sally, everything in my life after that bad thing has turned out to be better than I could have ever imagined.

Audible.com "Sally and the Magic River" (4:20:15).

Sally

This is where things get a little weird. That's not just a story for me. I actually had that accident and pertaining to your spin on dreams, if it hadn't been for my dream version of reality, I would not now exist and I would not now be talking to you. Just think about that!

Frank

Oh, don't worry. I've been thinking a lot about that lately. But Let's get back to the hard problem, Seth even asks the question, does an atom have a measurable level of consciousness? That's a crazy idea. But I've felt for a long time that the inanimate world has a lot more going for it than we give it credit. Sometimes I even believe it's only because I'm slightly smarter than they are that they've yet to kill me. I really have to be alert or I will get hurt! Have you ever watched a heavy picture frame launch itself in your direction for no good reason? I have. But, of course, I'm kidding when I think the atoms in the frame had anything to do with it, - - I think. Anyway, wouldn't it be great if there was a device that could measure consciousness much like a thermometer can measure temperature? Well there is such a device, and like I said, Seth's book is an eye-opener and a great listen for a walkjogrun. I'm also referencing one of Anil Seth's Ted talks and a NOVA documentary on the "hard" subject.

YouTube *"Your Brain Hallucinated Your Conscious Reality/AnilSeth"*, TED (17:01 LW).

YouTube *"Your Brain: Perception Deception/Full Documentary"*, (NOVA PBS (53:32 B).

The "Who is *Us*?"question is probably beyond our ability to answer in any complete way and there is way more to this subject than can be addressed in any book, much less this one. But isn't it a fascinating topic? I'm just going to stop here for others to ponder and maybe write their own AWTbook™ on the subject. However, I think I can almost guarantee that we, too, will be back with another AWTbook™ about something pertaining to the subject. It has been a blast getting to know you, Sally. Having written this book with you, I've come to realize that we humans really do have brains that can in effect represent two different people. Remember, the split brain experiments? They support this idea and now I've confirmed it for myself by finally realizing that you have been with me the whole time.

YouTube *"Curious: Split Brain"*, Thirteen (14:23 LW).

Sally

Don't you feel a little weird about all of this. I sure do.

No kidding, and here we go again. I was already to wrap this book up. But this happened and it just can't be a coincidence. I'm learning that this will almost always happen when one writes an AWTbook™. I just wrote the above crazy talk about simulations yesterday. This morning, I just *happened* to *justlisten* to a Klee Irwin video as I was *walkjogrunning*. He virtually took the words out of my mouth. He lines things up, just as I have tried to do here, to show the very high likelihood that we exist in a simulation, and one that exists within a simulation, that, exists within a simulation, etc., etc., etc.

YouTube "Klee Irwin - Are we living in a Simulation? - Part 4", Quantum Gravity (21:21 LW).

Sally
Wow! I wasn't expecting to see anything like this Klee Irwin video. I was just about to leave thinking your were completely bats. This is scary stuff, at the same time it's really exciting, mind boggling, and totally off-the-charts. What do we do now?

Frank
Good question, but before we go, you have to take a look at the next video. Marian Kerr, the girl in this video, is a crazy-bright actress who reminds me very much of you! I love her presentation.

YouTube "What is Reality? [Official Film}" Quantum Gravity (30:32 LW).

Sally
I see what you mean. She acts and even looks a little like me! It's amazing isn't it?These coincidences keep showing up. I can imagine myself being somewhat like her when I get a little older.

Frank
Hmm. That's odd, I wouldn't mind being her either when I grow up. You're 18 and already pretty grown up. As for me, I still feel like I'm 10 years old even though I'm 85. Based on that fact, I'm pretty sure I'll never "grow up". At any rate, I've got another surprise for you. I just came up with another video that connects to my nutty idea about the spherical elevator on which we stand.

YouTube *"Chapter 1-4: Rethinking General Relativity as 5 Dimensions of Physics – A unifying Theory of Gravity"*, Chris "The Brain" (1:03:21 B).

Sally

Oh my. Chris even uses the elevator analogy, and I really like his representation of 4-dimensions of space plus the dimension of time. His idea is that gravity can sort of be represented by what happens to a solid ball in jello where pressure on all surfaces of the ball can be made equal. He's worked the math out for this model and it does visually represent gravity better than the warping of space-time. But you and I have only *justlistened* to this one. We will have to spend a lot more time on it to better understand this new twist on reality. Maybe The Void acts like jello? Do you really think that idea could connect at all to your idea of the spherical elevator in a void?

Frank

Good question, Sally. More shall be revealed. But enough is enough. I've got to bring this "short" AWTbook™ to an end by ending it with what I think is most important

I love my life and I especially love the Earth Sets in the Morning and the Earth Rises in the Evening. As the sun disappears below the horizon, I can see and almost feel our planet's rotation as it blots out the sun. I'm always left in awe. I can't help but be grateful and applaud the Author for this gift. Is not life's show fantastic and incredibly entertaining? Does it not feel that we are but part time bit actors in a forever movie? Would not such a forever movie require infinitely fantastic entertainment? And for it to be a great forever movie, the infinitely entertaining good parts must be mixed with equally entertaining bad ones. You all know what I mean and why the bad parts are needed. If we had been protected and raised in a peaches-and-cream-only life we'd never make it, would we? When released into our survival of the fittest world, we'd perish, and even if we had been able to somehow survive we would surely be bored to death. I think the following movie says it all when It comes to entertainment and how its the major if not the only meaning of life. I'd just modify the title of the movie slightly from "The Good, The Bad and The Ugly" to "The Good, The Bad, The Ugly and the Lovely".

YouTube, *"The Good, The Bad, and The Ugly"*, YouTube Movies & Shows (2:58:00 LW).

YouTube *"10 Things You Didn't Know About the Day the Earth Stood Still"*, Minty Comedic Arts (18:19 B).

Sally

Life's entertaining alright. And I guess your idea does help explain why bad things happen. We clearly would not survive in this world without them and if everything was predictable we'd die from boredom. I don't know though. The bad stuff can sure go over the top. Mass shootings, wars, total global annihilation, I sure could do with a lot less of that. And, by the way, when was the last time you had a good belly laugh? Our world's entertainment is beginning to suck! But I have to ask. What does that last video about the Earth standing still have to do with anything?

Frank

I think "The Day the Earth Stood Still" goes directly to your point. It is a wonderful old movie, a wake-up-call-look at the possibility for us humans to have a glorious future. And, oh my, like you just pointed out, do we ever need help. We are in trouble. We hate each other. Our planet is in trouble and we are the ones at least partly responsible. That's the bad news but at the same time we are literally zooming into the possibility of a fantastic future! Kurzweil says we will be on our way to immortality by 2030! Musk says we'll be on our way to colonizing Mars by then. And I just heard about glassy associative polymers and the new idea that they work by the dissipation of free energy. Do you know what that means? I just realized that glassy associative polymers might be the means to finally invent that which has yet to be invented, THE FORCEFIELD of SCIENCE FICTION!! Yippee!!

Physical Review Letters *"Dynamics of Associative Polymers with High Density of Reversible Bonds"*, Phys. Rev. Lett. 130, 228101, May 2923.

YouTube *"Self-Siphoning Polymer"* Shorts (00:30 LW).

When I was 12, I began reading about such a future in the pulp science-fiction magazines of the 1950's. I have dreamed about living in a forcefield home, traveling to Mars and beyond, living long lives to be able to work sensibly with our miracle inventions of nuclear energy, AI, robots, etc. and be able to travel to the stars to meet our space-alien neighbors that surely must exist in this incredible Self-Assembling Universe of Ours. But to do all of this we will need to somehow prevent ourselves from going MAD. In other words we will need to stop going crazy and stop playing around with Mutual Assured Destruction. We are at that *"Gort Klaatu Barada Nikto"* moment in *"The Day the Earth Stood Still",* an old movie that in essence asks *"why* can't we all just get along?" We must figure that out soon or we will be toast, literally. To that point, I just remembered something that could be, oddly enough, hopeful. For many years I have been talking to whomever I could get to listen about the following idea. *"The only thing that can possibly bring humanity together would be an invasion from outer space".* We have a great example of the power behind that *coming together idea* from what happened in the United States with WWII. Essentially all of us in the United States and the United Kingdom got together to work against a common enemy! And to that point I just realized there is some good news to report. Without most of us realizing it we are currently experiencing such an invasion! Although it's not one of space-alien design it is, nonetheless, just such a potentially calamitous, dystopian invasion. We, the people of the world, MUST come together as a team to fight it or, as I just said, *we are toast.* It's the weather and she are attacking full-force as we speak! We either come together to fight this invasion or I think much of life on this planet will be done. As evidence, you may have noticed, we have already done a good job of exterminating it! So, I repeat. Don't you think we'd better at least begin to figure out how we can get along?

Sally

Of course I agree! You've really hit some nails on the head with that common enemy idea and given us *Us's* a lot of other heavy stuff to think about. But we can't do much about the *end of the world* subject right now. So, let's get back to the one at hand.

Who is Us remains a mystery, doesn't it? I still can't get over the fact that I exist on a large ball that spins about itself at a rate approaching 1000 miles per hour. Wherein I am held to the ball's surface no matter where I stand because my ball behaves like a directionless elevator that only can go up! Not only that my ball rotates about a freaking *star* at about 66,000 miles per hour! Give me a break! Oh, but I'm not done! My *star* rotates along with over 300-billion other stars that are gradually being sucked into a supermassive black hole!! What? That can't be right! Am I actually a space traveler traveling on the surface of a spaceship with a supermassive black hole as a companion!!?? Anyway, thankfully, my spaceship holds me fast to its surface as though I wore magical magnetic boots. And, get this. My spaceship has a see-though fuselage! I can actually see through it to see where I'm going with telescopic lenses that can see through my spaceship's skin that consists of a see-through, breathable, forcefield! And now here comes the best part. As I've been traveling through space with a galaxy having a black-hole at its center, I've been doing so at the rate of one million, three-hundred-thousand miles-per- hour! And, as I've been traveling at that terrific rate of speed I'm just beginning to realize the full significance of the fact that I have been self-made by atoms all of which began their creation 13.8 billion years ago and that are still creating me as we speak. From the fusion of protons, alpha-particles and other atomic nuclei in stars and by the stars very death in their calamitous, super-nova-last-breaths it became possible for me to pop into existence. I, along with you and all you other *Us's*, have been self-assembled by stardust! Now, my dear friend, I'm going to finally take that break and leave the rest to you.

Frank

That was fantastic. You really understand what I've been trying to say and get through my own thick head. You've said it all, Sally, but I will repeat because what you have said is, in my opinion, so profound it needs to be repeated. We are wondrous creatures who live with a vast diversity of other such creatures who are all held to the surface of a spherical spaceship that behaves like an elevator only going one way, up. Not only that, as you so beautifully described, Sally, our spaceship has a see-through, breathable fuselage.. As a result, we can see ourselves sailing through Our Universe, *and I'll just add again here, it's a universe that may very well be simulated from algorithms thought up by a simulator.* And let's not forget, this journey of ours through space is a dangerous one. Hardly any of us are meant to survive and, as far as I know, none of us *Us's* to date have yet figured out how to do that. Moreover, there are almost certainly other beings in Our Universe that are totally unlike us *Us's*. Have any of those beings figured out how to survive their own comings and goings? Have we already met them and not known it? Or will any of us *Us's* alive today be able to travel to the stars to meet them on their home planets? As ridiculous as that may sound, quite possibly so. At least Kurzweil thinks it possible since we seem to be on our way to living as long as we like, and we may even get there by 2030! Have there been other life forms on other planets that went MAD? There might have been many that had to deal with Mutual Assured Destruction and failed. In which case none of those life-forms now exist. Will we be just another example of such? That is very likely unless we somehow change our ways VERY SOON. I say all

8-billion plus of us *Us's* need to get on our knees and let Our Simulator know that we love our creation and can't wait to begin exploring Our Simulator's Simulation. I'm likely wrong about all this but why don't we all just get on our knees anyway and pray that we don't blow ourselves off the face of the Earth? What do we have to lose but the very existence of all life on this Spaceship we call Earth.

I have discovered that entertainment through the act of my own creativity is a wonderful way to lead a happy life. I feel aligned with my imagined Simulator when I do that. Writing this book is but one example. Sally and I encourage you to do one of your own. It's truly been an entertaining blast for us! It also has become clearer as we've written this book that entertainment is the very meaning of Our Simulation. What else is there to do if you happen to be a Simulator who lives forever, especially if you are a Simulator that time-shares with every Avatar in The Game? Seems like a good plan to me.

YouTube "The James Cameron's AVATAR game is surprisingly good", Keduit (33:47 LW).

Lastly, I think we may all need to get with it if we mean to survive our most creative, recent inventions, the AI's and Robots. What's your prayer? Mine? Aside from praying that we don't blow ourselves up or annihilate everything some other way, it's mostly just a humble *"Thank You"* filled with applause and gratitude as I watch the Earth's spin create Earth-sets in the morning and Earth-rises in the evening. I leave you with the following gift, a very recent look at exactly Where, When and How We are. Who knows? Maybe we will all be able to meet on Mars someday soon. But for now, check this out.

YouTube *"Unveiling the Universe: Is Everything Alive?"* Wisdom for Life (25:56).

Post-Script

I knew this would happen. Sally and I just discovered something that astonishes us, and made me feel more than a little under-educated. Thank goodness for Sally urging me to do some walkjogrunjustlistening this morning. Through all of our efforts here, it turns out that mostly what Sally and I accomplished in this book was to put a modern spin on a wheel that was already invented over 2000 years ago. So now, check this one out.

YouTube *"The Four Noble Truths of the Buddha Explained"*, Seek to Seeker (26:48 JL).

But I do like our modern drift on this matter. The first noble Buddhist truth is Dukkha. Apparently, this is what all of us *Us's* experience unless we are enlightened. As Sally and I understand it, Dukkha includes all the forms of suffering that one can think of, including the suffering experienced when good things come to an end. Clearly, Dukkha is the thing Sally and I refer to as entertainment and is the thing we regard to be the meaning of life in all its "good, bad, ugly and beautiful" forms. Buddhist Tanha, as we understand it, includes all of the "thirsts" we *Us's* have that lead to actual pain and to the disappointment that one experiences when good things come to an end. Also, as we understand it, Nirodah is the process of letting go *entirely* of all craving or thirsts by extinguishing one's *"self"*. Sally and I sort of do that by realizing the fact that I'm me and I'd be you, or that dog, plant or that rock if it weren't for the simple fact that we are we. "We", i.e., Sally and I, always exist as "me, myself and I" in some form. Marga is the 8-fold path to finding Buddhist Enlightenment. Sally and I know nothing about that process but *we* do have a sense for the Nirvana that an Enlightened Buddhist can, apparently, achieve. Sally and I feel that *we* at least begin to get to something like a state of Nirvana when we let ourselves go to "become", that is, to be *"one"* with everything. The first step on *our* path to enlightenment is to simply look deeply into the eyes of any living creature, especially those of a beloved. *We* also get very good results when *we* can do this with canines that we greet on *our* walkjogruns. *We* realize that *we* are not the Simulator, but the Simulator *is Us*. In other words *we* gain a closer connection to the feeling that *we* are one with the Simulator's Creations.

ALMOST THE END

Oops. Sally has taken leave but I, Frank, just came up with something so remarkable I must add it after the end. Sciepro.com is a website created by an amazing group that can show us the makings of *Us* creatures in beautiful, astonishing detail. I herewith add but a brief example.

YouTube *"Medically accurate 3d model of the Nervous System / Sciepro.com"*, #anatomy (00:46).

Also, in case you missed it, I might as well add this final video reference just so we have it handy for Sally and I to watch from time to time, and also so our readers can see it, especially if they have yet to understand just what I was saying to be the case when I said Our Universe is Simulated and Mathematically Designed by Algorithms. Here goes, for one last time. Given that we know that Consciousness exists, because we ourselves are examples of it, and given that we know that our seemingly solid material selves are but illusions of matter, because matter manifests itself from seething seas of self-annihilating virtual particles and antiparticles frothing about in the empty nothingness of the space that defines Our Universe, I think we know for sure that Our Universe is Simulated. And, if Our Universe is Simulated, it appears to be an *a priori* that Our Universe *must be Simulated by* a Conscious Simulator with an amazing set of algorithms that can simulate everything we experience. Anyway, based on all the evidence that we have accumulated through the writing of this book, that's what Sally and I think. However, and this is very important, we leave it to our readers to decide for themselves whether what is obvious to us is or is not a profound fact. The following video is long but well worth the time to set aside for a-watching. This video explains the basis for everything that we can see, feel, hear, taste and smell, without any mention of a Simulator, but interestingly enough, at the very end of the video that seems to be their bottom line. As I understand it, there *is* a Theory of Everything, even if nobody on Earth yet knows that theory in its entirety. Therefore, it is also clear that they hold the truth that Our Universe is Mathematical and that there exists an algorithm that could be designed to Simulate and Run Everything.

And check out the title of this Chaos Video and have a look. You'll be glad you did. Jim is a star at explaining this complex stuff.

YouTube *"Chaos Theory: The Science Behind the Miracle of Intelligent Life / Jim Al-Khalili",* Doc of the Day (50:26).

And now you know why Sally and I think we may exist in a system of forever that can't have an end or a beginning and now know with both of us the answer to the question.

We is *Us.*

Who is I?

In addition to our plan to make a movie based on our book, *Sally and the Magic River,* Sally and I have seven other literary works that one might use for that purpose or use to make some amusing, AI-generated, video experiences. For example the five-volume set, The *Amazing Illustrated Word Game Memory Books,* presents an odd-ball mix of over 100 illustrated-stories which could be used for cinematic inspiration. One might wonder, "How and why did you do that?" I worked on it for ten years, that's how! As to why, here's my unbelievably ridiculous answer. My wife and a friend of hers invited me to play Scrabble and then proceeded to render me a babbling idiot. I couldn't even unscramble a few little words from a random rack of seven letters! So, with that humiliating defeat, of course I did what anybody would do. I went to work for 10-years creating short-stories and drawings that could help me remember and unscramble seven-letter, bonus words. What else was there to do? Anyone would do that, right? And, proof to the pudding, my efforts paid off! By the time I completed the series I had risen to the level of Half-Idiot Scrabble-Player! And there were other benefits. I learned to draw some pretty cool pictures. For example, just imagine a poor girl named "ENTASIA" having spasms as she stands nearly naked in the snow but respectably dressed only in a man's tie and "TAENIAS", aka, Greek head-bands. These two bonus words are what one gets when one adds an additional letter-A and unscrambles the six remaining letters of INEAST. Isn't that amazing? One gets a 50-point, Scrabble-Bonus BINGO when anyone can make such a play! The pictures and stories that I put together work because I used respectful humor and pictures (really, I went out of my way to make it so for everyone) and the Lorayne and Lucas *Memory Book* method throughout the *Amazing-Illustrated* series. For instance in the above example a man's *tie is a* visual cue for the added letter A. I made such a cue for all 26 letters of the alphabet in order to win SCRABBLE games. And, because the method worked so well, I went on to create a total of 118 such visual alpha-numeric cues just so I could count myself to sleep with surrogate sheep. To wit, as I drift off, I count the atoms in the periodic table of known-to-exist elements and see 118 atomic-numbers dancing in my head instead of sheep or sugar plums. For example, I specifically see a *dove playing an organ* to help bring up the name, Oganesson. Oganesson is the name scientists gave to the nucleus of the atom they made to contain 118 protons. That nucleus only lasts for milliseconds. Why a *dove*? It's my code for number 118.

As for Scrabble, I'm now very happy with my game. I just won three straight from my in-house Scrabble nemesis. Since the two of us are now almost perfectly-matched competitors, we often enjoy egging each other on. For instance, my wife likes to sweetly say to me when she wins, "You had a lousy public education didn't you?" I retort, "True enough, but I made up for it, didn't I? And now my dear-one *you* are really in for it!" Needless to say, we both love Scrabble.

I'm also the author and co-author of over 50 scientific papers and patents, one patent is for the world's first and only successfully used, genetically-engineered, organic insecticide. Another cool patent is for an encapsulated form of interferon-gamma which can increase the cytokine's activity by 1000-fold, wherein low-dose administrations eliminate the cytokine's bad side effects. This is a product that has the potential to stop many viral infections in both humans and domestic animals. However, it remains shelved by the chemical company who bought our Company over twenty years ago and has since left the invention to be completely ignored by the scientific community. I retired in year-2000 and, regrettably, also have left the invention for others to pursue. It needs to be, don't you think?

As for my educational background, I was a low grade C and B student until I graduated in 1956 from high-school. Then everything changed. I decided to begin in earnest the advancement of my wannabe musical career at a Junior College. It was there that I first surprised myself by blowing the curve in science classes. I still can't believe how I managed to excel at these tough courses while I was also vigorously studying music with great interest and focus. Wherever did I find the time? I even composed and had performed an original, four-part work for brass-choir. I was in the audience for that one just to be able to hear my own work being played. I impressed myself so much I can still hear the music and the applause of the standing ovation resounding in my head. I mention this event because it explains a lot about my quirky, one-off behavior.

My father died when I was ten and I think it left me not relying on anyone, even myself. I was a loner and that one-off performance is a good example of many future one-offs. I now understand this behavior as having been an early example of me scaring my little self with success. My Dad was successful and then he died. I don't want to do that. So with success, until very recently, it seems I just cut and ran! I've come to realize that a lot of people may scare themselves this way because of some tragic thing that happens to them at an early age. Just watch American Idol or Survivor and you'll see what I mean where people seem to battle with themselves to contend with this pervasive trait. Clearly, for some reason many of us *Us's* don't give ourselves much credit. We scare our little selves and then give up on our dreams at the first sign of success. Again, I ask, why is that? I'm sure fear of success is often involved but it also might have something to do with what comes with a distracted wandering mind. I definitely have one of those as well.

At the University of Arizona I enthusiastically continued to study music. I played Bach on the pipe organ, sang in a church choir, played in a dance band, played solo piano at a club, composed music and orchestrated my own work for a full symphony orchestra. That was all great fun but I kept telling myself I wasn't good enough, and required science courses were once again attracting my attention. My mind began to wander and I started to wonder if I was following the right path. Clearly, my ability to make enough money to survive with music alone wasn't going to make it. I thought I might eventually be good enough to survive but only if I supplemented my income with something more. At any rate, it was necessary for me to do additional things to get through school. So, while I worked at school studies, I drove a Volkswagen Minibus to transport checks from bank-branches to the home office. Mind you, this was the old days. Banks don't do that anymore. Everything is digital and sent electronically. I also counted checks, balanced books, ran the bank's large ledger machines and had wonderful, wild-west rubber-band fights during breaks. Obviously, there were signs I wasn't going make a great banker. I wonder, can anyone still have fun at a bank? I'm thinking not. I even hauled concrete pipe to worksites for a concrete-pipe company but that didn't last long. I was no good at it. I soon got fired when I broke too many pipes. Nope, that wasn't going to supplement my work as a musician either. In the meantime, I continued to do well with required science classes and the bells for a career in science began to chime. In 1962 I decided to go for a graduate degree. Still, my plan was just to get a masters degree to help me get a job as a lab-tech or something that could support my real life as a musician. But as I worked on that degree I found I got very excited by the idea of research, an idea that became especially true when I made my first "great" discovery!

I don't think anybody but me, and now you dear reader, have any idea or even care that I was arguably the first person on Earth to discover and write about the feeding habits and unique survival skills of a multi-nucleate, single-cell creature who lives in extremely hot conditions, can grow to be larger than the inside of a bathtub, can move about totally naked and exposed, and then, out of its many branches of protoplasmic goo can recreate itself into a field of beautiful, teed-up, golf-ball-like structures. Sounds creepy, delightful and impossible to imagine, does it not? As luck would have it, under the wonderful guidance of my thesis adviser, Dr. Adelaide Evenson, I discovered a slime mold called *Physarum pusillum* living happily under seemingly impossible conditions in a boiling hot dessert. I learned how this very fascinating, active and hungry creature ate. I could actually see, with my very own two eyes aided by a wonderful microscope, its food, a red yeast, rapidly circulating through Its gelatinous, transparent body. Moreover this *oughttobedelicate* creature was somehow able to survive in the 120-degree heat of Somerton, Arizona. I discovered how both the yeast and the slime mold were able to support each other in the bark and sap of a Mesquite tree growing in the desert that commenced just a few feet from my parent's backyard. Interesting, isn't it? And now, see for yourself some of the fun and great discoveries that people are currently experiencing with this "Thing". The link below shows why one can get very serious about working with these beautiful yet creepy-looking alien beings. Today's "true" slime mold of choice is *Physarum polycephalum* and as you will see in the link after my thesis record, *P. polycephalum* is all the rage. However, except for its yellow color and its choice to live in a lush forest, the latter version of *Physarum* is very similar to the one I studied.

MS Thesis *"A Study of the Nutritional and Environmental Factors Influencing the Growth and Maintenance of a Desert Strain of Physarum pusillum martin"*, Frank H. Gaertner,1962, University of Arizona.

No wonder I got so excited and sucked into science instead of music. I was having a blast. Of course, I didn't stay with slime molds. I'm a one-off and nobody encouraged me to stick with it. I was also just happy to get my degree and get the heck out of the sweltering heat. By the way, Tucson is paradise compared to Yuma. I was a High School Yuma Criminal for four years so I guess I should know. As you can tell, I don't much like the heat which explains my love for San Diego's marine layer. God only knows what would have happened if I had continued my work with slime molds. I might have shriveled from heat but based on all the current excitement I'd probably be famous as a pioneer in the field. For sure my life would have been totally altered in time, place and circumstance. That, too, is interesting, isn't it? I get kind of jealous when I see this video. See for yourself the fun people are now having with the yellow-version of my Master's Thesis.

YouTube, *"Breakthrough: The Slime Minder"*, SciFri (10:15).

As I was about to begin searching for a lab-tech job, a man who would soon become my new thesis advisor showed up in 1962 to give a lecture at the University of Arizona on one of the very first proteins to be sequenced to reveal a protein's complete amino acid structure. "What's an amino acid?" I vividly remember being embarrassed when I actually posed that question at this man's seminar. But why *would* I know anything about amino acids? I was a music major for goodness sake! Needless to say, I was impressed by this austere guy with a heavy Austrian accent who just presented a huge scientific breakthrough. The first complete sequence of a large protein like flagellin, i.e., the self-assembling protein of the little whip-like structures that bacteria use to move, was a really big deal. So, to say the least, I was impressed. But surprisingly the seminar presenter must also have been impressed, and that was with me. But clearly it wasn't with my knowledge of science. Rather, I think he was impressed with my thesis and the nerve it took for me to ask a stupid question. At any rate he actually invited me *on the spot* to come to Purdue to get a PhD! And I was going to get to be paid to do that with one of his new training grants! What?! That's right us upper-degree students in those days really had it good. However, that good thing was not to last. Soon after I graduated with a PhD the political party in charge at the time said taxes were too complicated and the government was too occupied helping people get an education. *No they weren't and no the government wasn't!* In those days I could do my own taxes. Instead of simplification, taxes just changed in a way to benefit the rich and famous. In addition training grants vanished and so did a lot of other things, including jobs that moved off shore. But no worries, the latter brilliant move was thanks to a different political party that should have known better. Now students can get loans and live in a tent. Grrrr!

In 1965 at Purdue University I was once again onto something that had world-shattering potential. I was the first to accomplish and write about how the *in vitro* synthesis of proteins could be accomplished at high temperature using the DNA polymerase and other protein synthesizing machinery isolated from thermophilic bacteria. Note, what I did in 1965 turned out to be a necessary part of the work that led to a major breakthrough that's now used by CSI. This is long before PCR became a major part of binge-watching-TV and the household word of a pandemic. But did I follow it up? No. Once more I scared my little self. Once more, god only knows what would have

97

happened if I had gone on in the 1960s to isolate the thermophilic DNA-polymerase that's now used in PCR. BTW, PCR wasn't invented until 1985. A surfer dude named Kary Mullis came up with the idea for which he got the Nobel Prize! (When I was looking up Kary to make sure his name was spelled with a K, *I just discovered with sadness and a shock that Kary died at age 74 in 2019! That seems so young to me. I'm now 85 and still going strong.*

There's my proof. Success can be dangerous. I'm still here because I avoided major breakthroughs? Don't be stupid Frank! But I guess that's my story so I'll stick to it because the above is just one of the major breakthroughs that "I can be very happy to have let slip through my fingers". The next breakthrough didn't go anywhere either due to my one-off "malady". I was doing research that was at the very frontier of molecular genetics itself! The work I did constructing the world's first successful *in vitro* synthesis of a functional protein has recently become a very big deal. The two publications below show how my work was done at a *very* early date.

PhD Thesis *"The Cell-Free Synthesis of Flagellin"*, Frank H Gaertner, 1965, Purdue University.

Giornale Botanico Italiano *"The Synthesis In Vitro of Flagellin by a Cell-Free Extract from Bacillus pumilus"*, F. Sala, F. Gaertner and H. Koffler, 1968.

The following video link will reveal the fun and importance that *cell-free protein* synthesis has had in its current use nearly 60 years later! At that time long ago I was once again a loner and the only person late at night in my dismal lab in the basement of Purdue's Lilly Hall doing the work. I had no help. Nevertheless, I was determined to succeed with a task that seemed impossible. Really, at that time nobody had yet been successful at synthesizing a large *functional* protein in a test tube! Even though my work had been a spectacular success, and even though that success had been repeated by another researcher to confirm it, that success wasn't followed up. Why? Because I cut and ran and my advisor naturally became more interested in his new job as President of the University of Arizona. And I, in my usual unnatural way, proceeded to do an entirely different line of research at the University of California in San Diego. Once again, I had literally terrified my little self with success. My results are shown in my 1965 thesis and

Dr Francesco Sala's confirmations are shown in the referenced 1968 publication. But here's the cool thing. You can see in the YouTube link below some of the very important work now being done using molecular genetics and the method of *Cell-Free Protein Synthesis.*

YouTube *"An Introduction to Cell-Free Protein Expression"*, The Scientist Creative Services Division (3:19).

At UC San Diego I again found myself working alone investigating something entirely different, a biochemical pathway wherein enzymes are tightly bound together in linearly arranged chains called multienzyme complexes. How did such enzyme complexes work and did they facilitate an overall increase in the efficiency of converting long substrate/product chains into end-products? At least here, I took a shot at sticking with it. I continued this work in 1969 as a University of Tennessee Professor at a graduate school in Oak Ridge National Laboratory. This time I had excellent technical assistance and they helped me discover the amazing catalytic properties of five enzymes covalently linked together to operate as if the five were operating as a single, five-step, catalytic protein. This work, too, had futuristic possibilities. For example, covalently linked multienzyme pathways can now be easily created to produce new drugs and other valuable products. From DNA analysis and the creation of functional three-dimensional sequences of enzymes using artificial intelligence one can conceive and produce any such combination for the purpose of efficiently and inexpensively producing valuable complex end-products. But, did I follow that predictable future up? Nope.

In 1981 at the Salk Institute's SIBIA Corporation and with the substantial help of two excellent technicians, we isolated the *his*-3 gene from *Pichia* yeast for use in SIBIA's new yeast vector which is currently in worldwide use for protein expression. But, again, I quickly moved on to do something entirely different.

In 1982 a colleague, a business associate, a venture capitalist and I co-founded Mycogen Corporation. As Director of that company's Molecular Genetics department, as described at the beginning of this "Who is I" compilation, I headed up a team of scientists that made the world's first and only genetically-engineered, organic insecticide. That led to our company being acquired wherein I was able to stay on to assemble yet another team where we created the interferon-gamma-ARCs invention referred to earlier. So, that's it. That's my "sad" story of how I created for myself a miracle life that I would not want to trade with anyone. Moral of the story? Ya pays yer money and ya takes yer chances. Life is an amazing trip full of dangerous entertainment in which one gets to make decisions that take one into the unknown. It's crazy exciting, is it not?

To end this self-serving tirade, you might find it as surprising as I do that I am still a wannabe musician who is frequently a top ten in SD's ReverbNation. I'd like to think, but I'm probably wrong, that my recognition on ReverbNation is mostly due to my orchestration of Bach's G# minor Fugue. Sorry Bach but I just wanted to show the world how wonderful it would be if all 24 of your preludes and fugues in your Well Tempered Clavier were orchestrated and featured in live performance by demonstrating how it might sound with my amazing Peavy DPM 3SE Synthesizer. As for the importance of that G# minor fugue, I brazenly nicknamed it *Rivulets* for use as *Sally's Theme* in the audible version of *Sally and the Magic River*. Of all my *wannabe-great-before-I-croak* accomplishments, I'm most proud of *Sally and the Magic River*, especially its Audible version, cf, *sallysmagicriver.com*, and my OSAU series, especially this volume, the *OSAU-3, Who is Us edition*. Sally and I really hope you really, really, like it, *really*.

Important Afterthoughts

I don't need to cut off my arm to know. I've been flirting with this answer for a long time. And, now, based on an Alan Watts lecture, I think I have really settled on just Who *I is*. Although the answer is obvious, *it's nothing that any of us can actually talk about*. Our reality is beyond words. The only useful things we can say are things like *"This flax seed weighs three pounds"* or *"The only reason I am not you is that I am me."*

YouTube "30 min of PURE GENIUS-Alan Watts on The Gateless Gate", Anima Creativa (30:11).

And, just in case you missed it earlier, the following video-link features a girl who looks and sounds very much like Sally. She is also similar to my Sally who is much smarter than my Frank. She explains reality using the golden ratio, crystals simulated by tetrahedral Planck-length pixilation, matrix algebra and consciousness. As noted before in OSAU-2, C&I&L&E=mc², we humans are getting very close to unraveling the truth and meaning of reality. Klee Irwin and others are homing in on a new Theory of Everything based on blending Einstein's relativity with quantum mechanics. Their developing "everything" equation uses the golden ratio as found in black holes and uses it as a key that will ultimately reveal our existence to be a simulation which is but one in a series of such simulations. The idea is that endless successions of simulations are created by hierarchies of conscious, super-quantum-computing entities that use algorithms consisting in part of the following:

1. The Golden-Ratio that's hidden in circumscribed equilateral triangles and black holes,
2. Tetrahedral E8-crystal pixilation.
3. Nondeterministic causality loops.

Klee and others are still working on this, and Sally and I are still trying to understand it but the above is what we think we understand. We sort of like Irwin's "Entities" better than words like Gods or Game Masters, but there is no doubt, at least not in the superimposed minds of Sally and Frank, that Our Universe's Entity enjoys and needs, as much as all the rest of us *Us's* do, the infinite entertainment-value found in the full-participation love/hate battles that take place in Our Good, Bad, Ugly, Love, Hate and Beautiful Universe. We are all in this together created by Our Entity's algorithms and imaging of all things. You can't deny it. Our Universe consists of nothing but great entertainment. So, we might as well enjoy it and dispense as best we can of the hate. And, as for pain, "It only hurts for a little while. That's what they tell me. That's what they say". Emanuel Kant said, "I think, therefore I am." Meaning in my interpretation, "I think, therefore I hallucinate myself into existence, or I think, therefore another entity hallucinates me into existence." Which ever it is, aside from faith, I *truly* know only one thing, *I am.* So the question remains. Who am I? Aside from all the nonscientific fiddle-dee-dee, I know this. I am a thing *made by* forcefields that show up in me as an atomic workforce of assemblers over seven-octillion strong, a number that happens to be more than all the stars thought to be in Our Visible Universe. And, as I've already said, these pico-scale workers assemble nano-scale molecules that reassemble and repair parts of me each and every day much as a Lego Set might do it when magically brought to life to reassemble parts of a broken Lego Robot. Really? I can't help it. I've got to say it again, "You have got to be kidding me!" But, and here's the thing, all the scientific evidence says that nobody is kidding here. The reality of Our Self-Assembling Universe is obvious to see, but only if one looks hard enough and begins their looking by throwing a block of atomic number-11 metal into water and *watchlistens* to two slugs mating as they dangle from a silvery slime-rope. Having done that one should now be fully ready to give a watchlisten to the following video reference with a "Hey Siri! YouTube! What is Reality? Official Film".

YouTube "What is Reality? Official Film", Quantum Gravity Research (30:19).

AWTbook™ References

Surprise Bonuses just for us *Us's* and anyone else who might happen to glance at the following "note added in proof" reference list:

105. YouTube *"Beyond the Atom: Incredible Plunge into the Heart of Matter"*, Wondody / The World of Odysseys (1:33;18).---

106. Audible "The Song of the Cell", Siddhartha Mukherjeej (10:14:52).--------------------------

107. YouTube "Physicist Observed for the First Time How Reality Works / Nobel Prize in Physics 2023 Explained", EXOPLANET-Sci (12:32).--

Holy Attoseconds! We can now SEE electrons and we can see them in motion!! So, yes. Heisenberg was wrong. But are electrons ever quick. An attosecond is just one quintillionth of a second!! (That's 0.000000000000000001 of a second.)

www.ingramcontent.com/pod-product-compliance
Lightning Source LLC
Chambersburg PA
CBHW052342210326
41597CB00037B/6233